**Advanced Courses in Mathematics
CRM Barcelona**

Centre de Recerca Matemàtica

Managing Editor:
Manuel Castellet

Alexei Myasnikov
Vladimir Shpilrain
Alexander Ushakov

# Group–based
# Cryptography

Birkhäuser Verlag
Basel · Boston · Berlin

Authors:

Alexei Myasnikov
Department of Mathematics and Statistics
McGill University
Montreal, Quebec H3A 2K6
Canada
e-mail: amiasnikov@gmail.com

Vladimir Shpilrain
Department of Mathematics
The City College of New York
New York, NY 10031
USA
e-mail: shpil@groups.sci.ccny.cuny.edu

Alexander Ushakov
Department of Mathematics
Stevens Institute of Technology
Hoboken, NJ 07030
USA
e-mail: sasha.ushakov@gmail.com

2000 Mathematical Subject Classification: 11T71, 20Exx, 20Fxx, 20Hxx, 20P05, 60B15, 68P25, 94A60, 94A62

Library of Congress Control Number: 2008927085

Bibliografische Information Der Deutschen Bibliothek
Die Deutsche Bibliothek verzeichnet diese Publikation in der Deutschen Nationalbibliografie; detaillierte
bibliografische Daten sind im Internet über <http://dnb.ddb.de> abrufbar.

ISBN 978-3-7643-8826-3 Birkhäuser Verlag, Basel · Boston · Berlin

© 2008 Birkhäuser Verlag, P.O. Box 133, CH-4010 Basel, Switzerland
Part of Springer Science+Business Media
Cover design: Micha Lotrovsky, 4106 Therwil, Switzerland
Printed on acid-free paper produced from chlorine-free pulp. TCF ∞

ISBN 978-3-7643-8826-3

ISBN 978-3-7643-8827-0 (eBook)

9 8 7 6 5 4 3 2 1

www.birkhauser.ch

*To our parents*

# Contents

## II  Non-commutative Cryptography                                        35

# Preface

This book is about relations between three different areas of mathematics and theoretical computer science: combinatorial group theory, cryptography, and complexity theory. We explore how non-commutative (infinite) groups, which are typically studied in combinatorial group theory, can be used in *public key cryptography*. We also show that there is a remarkable feedback from cryptography to combinatorial group theory because some of the problems motivated by cryptography appear to be new to group theory, and they open many interesting research avenues within group theory. Then, we employ complexity theory, notably *generic case complexity* of algorithms, for cryptanalysis of various cryptographic protocols based on infinite groups. We also use the ideas and machinery from the theory of generic case complexity to study *asymptotically dominant properties* of some infinite groups that have been used in public key cryptography so far. It turns out that for a relevant cryptographic scheme to be secure, it is essential that keys are selected from a "very small" (relative to the whole group, say) subset rather than from the whole group. Detecting these subsets ("black holes") for a particular cryptographic scheme is usually a very challenging problem, but it holds the key to creating secure cryptographic primitives based on infinite non-commutative groups.

The book is based on lecture notes for the Advanced Course on Group-Based Cryptography held at the CRM, Barcelona in May 2007. It is a great pleasure for us to thank Manuel Castellet, the Honorary Director of the CRM, for supporting the idea of this Advanced Course. We are also grateful to the current CRM Director, Joaquim Bruna, and to the friendly CRM staff, especially Mrs. N. Portet and Mrs. N. Hernández, for their help in running the Advanced Course and in preparing the lecture notes.

It is also a pleasure for us to thank our colleagues who have directly or indirectly contributed to this book. Our special thanks go to E. Ventura who was the coordinator of the Advanced Course on Group-Based Cryptography at the CRM. We would also like to thank M. Anshel, M. Elder, B. Fine, R. Gilman, D. Grigoriev, Yu. Gurevich, Y. Kurt, A. D. Miasnikov, D. Osin, S. Radomirovic, G. Rosenberger, T. Riley, V. Roman'kov, A. Rybalov, R. Steinwandt, B. Tsaban, G. Zapata for numerous helpful comments and insightful discussions.

We are also grateful to our home institutions, McGill University, the City

College of New York, and Stevens Institute of Technology for a stimulating research environment. A. G. Myasnikov and V. Shpilrain acknowledge support by the NSF grant DMS-0405105 during their work on this book. A. G. Myasnikov was also supported by an NSERC grant, and V. Shpilrain was also supported by an NSA grant.

Alexei Myasnikov                                                              Montreal,
Vladimir Shpilrain                                                            New York
Alexander Ushakov

# Introduction

The object of this book is twofold. First, we explore how non-commutative groups which are typically studied in combinatorial group theory can be used in *public key cryptography*. Second, we show that there is a remarkable feedback from cryptography to combinatorial group theory because some of the problems motivated by cryptography appear to be new to group theory, and they open many interesting research avenues within group theory.

We reiterate that our focus in this book is on public key (or asymmetric) cryptography. Standard (or symmetric) cryptography generally uses a single key which allows both for the encryption and decryption of messages. This form of cryptography is usually referred to as symmetric key cryptography because the same algorithm or procedure or key is used not only to encode a message but also to decode that message. The key being used then is necessarily private and known only to the two parties involved in communication. This method for transmission of messages was basically the only way until 1976 when W. Diffie and M. Hellman introduced an ingenious new way of transmitting information, which has led to what is now known as public key cryptography. The basic idea is quite simple. It involves the use of a so-called one-way function $f$ to encrypt messages. Very informally, a one-way function $f$ is a function such that it is easy to compute the value of $f(x)$ for each argument $x$ in the domain of $f$, but it is very hard to compute the value of $f^{-1}(y)$ for "most" $y$ in the range of $f$. The most celebrated one-way function, due to Rivest, Shamir and Adleman, gives rise to the protocol called RSA, which is the most common public key cryptosystem in use today. It is employed for instance in the browsers Netscape and Internet Explorer. Thus it plays a critical and increasingly important role in all manner of secure electronic communication and transactions that use the Internet. It depends in its efficacy, as do many other cryptosystems, on the complexity of finite abelian (or commutative) groups. Such algebraic structures are very special examples of *finitely generated groups*. Finitely generated groups have been intensively studied for over 150 years and they exhibit extraordinary complexity. Although the security of the Internet does not appear to be threatened at this time because of the weaknesses of the existing protocols such as RSA, it seems prudent to explore possible enhancements and replacements of such protocols which depend on finite abelian groups. This is the basic objective of this book.

The idea of using the complexity of infinite nonabelian groups in cryptography goes back to Magyarik and Wagner [97] who in 1985 devised a public-key protocol based on the unsolvability of the word problem for finitely presented groups (or so they thought). Their protocol now looks somewhat naive, but it was pioneering. More recently, there has been an increased interest in applications of nonabelian group theory to cryptography (see for example [1, 84, 129]). Most suggested protocols are based on *search problems* which are variants of more traditional *decision problems* of combinatorial group theory. Protocols based on search problems fit in with the general paradigm of a public-key protocol based on a one-way function. We therefore dub the relevant area of cryptography *canonical cryptography* and explore it in Chapter 4 of our book.

On the other hand, employing decision problems in public key cryptography allows one to depart from the canonical paradigm and construct cryptographic protocols with new properties, impossible in the canonical model. In particular, such protocols can be secure against some "brute force" attacks by a computationally unbounded adversary. There is a price to pay for that, but the price is reasonable: a legitimate receiver decrypts correctly with probability that can be made arbitrarily close to 1, but not equal to 1. We discuss this and some other new ideas in Chapter 11.

There were also attempts, so far rather isolated, to provide a rigorous mathematical justification of security for protocols based on infinite groups, as an alternative to the security model known as *semantic security* [50], which is widely accepted in the "finite case". It turns out, not surprisingly, that to introduce such a model one would need to define a suitable probability measure on a given infinite group. This delicate problem has been addressed in [17, 16, 89] for some classes of groups, but this is just the beginning of the work required to build a solid mathematical foundation for assessing security of cryptosystems based on infinite groups. Another, related, area of research studies *generic* behavior of infinite groups with respect to various properties (see [75] and its references). It is becoming clear now that, as far as security of a cryptographic protocol is concerned, the appropriate measure of computational hardness of a group-theoretic problem in the core of such a cryptographic protocol should take into account the "generic" case of the problem, as opposed to the worst case or average case traditionally studied in mathematics and theoretical computer science. Generic case performance of various algorithms on groups has been studied in [75, 77], [78], and many other papers. It is the focus of Part III of this book.

We have to make a disclaimer though that we do *not* address here security properties (e.g., semantic security) that are typically considered in "traditional" cryptography. They are extensively treated in cryptographic literature; here we single out a forthcoming monograph [51] because it also studies how group theory may be used in cryptography, but the focus there is quite different from ours; in particular, the authors of [51] do not consider infinite groups, but they do study "traditional" security properties thoroughly.

In the concluding Part IV of our book, we use the ideas and machinery from Part III to study *asymptotically dominant properties* of some infinite groups that have been used in public key cryptography so far. Informally, the point is that "most" elements, or tuples of elements, or subgroups, or whatever, of a given group have some "smooth" properties which makes them unfit for being used (as private or public keys, say) in a cryptographic scheme. Therefore, for a relevant cryptographic scheme to be secure, it is essential that keys are actually selected from a "very small" (relative to the whole group, say) subset rather than from the whole group. Detecting these subsets ("black holes") for a particular cryptographic scheme is usually a very challenging problem, but it holds the key to creating secure cryptographic primitives based on infinite nonabelian groups.

# Part I

# Background on Groups, Complexity, and Cryptography

In this part of the book we give necessary background on public key cryptography, combinatorial group theory, and computational complexity. This background is, of course, very limited and is tailored to our needs in subsequent parts of the book.

Public key cryptography is a relatively young area, but it has been very active since its official beginning in 1976. By now there are a great many directions of research within this area. We do not survey these directions in the present book; instead, we focus on a couple of the most basic, fundamental areas within public key cryptography, namely on key establishment, encryption, and authentication.

Combinatorial group theory, by contrast, is a rather old (over 100 years old) and established area of mathematics. Since in this book we use group theory in connection with cryptography, it is not surprising that our focus here is on algorithmic problems. Thus, in this part of the book we give background on several algorithmic problems, some of them classical (known as *Dehn's problems*), others relatively new; some of them, in fact, take their origin in cryptography.

Probably nobody doubts the importance of complexity theory. This area is younger than combinatorial group theory, but older than public key cryptography, and it is hard to name an area of mathematics or theoretical computer science that would have more applications these days than complexity theory does. In this part of the book, we give some background on foundations of computability and complexity: Turing machines, stratification, and complexity classes.

# Chapter 1

# Background on Public Key Cryptography

In this chapter we describe, very briefly, some classical concepts and cryptographic primitives that were the inspiration behind new, "non-commutative", primitives discussed in our Chapter 4. It is not our goal here to give a comprehensive survey of all or even of the most popular public key cryptographic primitives in use today, but just of those relevant to the main theme of our book, which is using non-commutative groups in cryptography. In particular, we leave out RSA, the most common public key cryptosystem in use today, because its mechanism is based on Euler's generalization of Fermat's little theorem, an elementary fact from number theory that does not yet seem to have any analogs in non-commutative group theory.

For a comprehensive survey of "commutative" cryptographic primitives, we refer the reader to numerous monographs on the subject, e.g., [45], [100], [134].

Here we discuss some basic concepts very briefly, without giving formal definitions, but emphasizing intuitive ideas instead.

First of all, there is a fundamental difference between public key (or *asymmetric*) cryptographic primitives introduced in 1976 [27] and *symmetric* ciphers that had been in use since Caesar (or even longer). In a symmetric cipher, knowledge of the decryption key is equivalent to, or often exactly equal to, knowledge of the encryption key. This implies that two communicating parties need to have an agreement on a shared secret before they engage in communication through an open channel.

By contrast, knowledge of encryption and decryption keys for asymmetric ciphers are not equivalent (by any feasible computation). For example, the decryption key might be kept secret, while the encryption key is made public, allowing many different people to encrypt, but only one person to decrypt. Needless to say, this kind of arrangement is of paramount importance for e-commerce, in particular for electronic shopping or banking, when no pre-existing shared secret is possible.

In the core of any public key cryptographic primitive there is an alleged practical irreversibility of some process, usually referred to as a *trapdoor* or a *one-way function*. For example, the RSA cryptosystem uses the fact that, while it is not hard to compute the product of two large primes, to *factor* a very large integer into its prime factors seems to be very hard. Another, perhaps even more intuitively obvious, example is that of the function $f(x) = x^2$. It is rather easy to compute in many reasonable (semi)groups, but the inverse function $\sqrt{x}$ is much less friendly. This fact is exploited in Rabin's cryptosystem, with the multiplicative semigroup of $\mathbb{Z}_n$ ($n$ composite) as the platform. For a rigorous definition of a one-way function we refer the reader to [135]; here we just say that there should be an efficient (which usually means polynomial-time with respect to the complexity of an input) way to compute this function, but no visible (probabilistic) polynomial time algorithm for computing the inverse function on "most" inputs. The meaning of "most" is made more precise in Part III of the present book.

At the time of this writing, the prevailing tendency in public key cryptography is to go wider rather than deeper. Applications of public key cryptography now include digital signatures, authentication protocols, multiparty secure computation, etc., etc. In this book however the focus is on cryptographic *primitives* and, in particular, on the ways for two parties (traditionally called Alice and Bob) to establish a common secret key without any prior arrangement. We call relevant procedures *key establishment protocols*. We note that, once a common secret key is established, Alice and Bob are in the realm of symmetric cryptography which has obvious advantages; in particular, encryption and decryption can be made very efficient once the parties have a shared secret. In the next section, we show a universal way of arranging encryption based on a common secret key. As any universal procedure, it is far from being perfect, but it is useful to keep in mind.

## 1.1  From key establishment to encryption

Suppose Alice and Bob share a secret key $K$, which is an element of a set $\mathcal{S}$ (usually called the *key space*).

Let $H : \mathcal{S} \to \{0,1\}^n$ be any (public) function from the set $\mathcal{S}$ to the set of bit strings of length $n$. It is reasonable to have $n$ sufficiently large, say, at least $\log_2 |\mathcal{S}|$ if $\mathcal{S}$ is finite, or whatever your computer can afford if $\mathcal{S}$ is infinite. Such functions are sometimes called *hash functions*. In other situations hash functions are used as compact representations, or digital fingerprints, of data and to provide message integrity.

**Encryption:**  Bob encrypts his message $m \in \{0,1\}^n$ as

$$E(m) = m \oplus H(K),$$

where $\oplus$ is addition modulo 2.

**Decryption:** Alice computes:

$$(m \oplus H(K)) \oplus H(K) = m \oplus (H(K) \oplus H(K)) = m,$$

thus recovering the message $m$.

Note that this encryption has an *expansion factor* of 1, i.e., the encryption of a message is as long as the message itself. This is quite good, especially compared to the encryption in our Section 6.1, say, where the expansion factor is on the order of hundreds; this is the price one has to pay for security against a computationally superior adversary.

## 1.2 The Diffie-Hellman key establishment

It is rare that the beginning of a whole new area of science can be traced back to one particular paper. This is the case with public key cryptography; it started with the seminal paper [27]. We quote from Wikipedia: "Diffie-Hellman key agreement was invented in 1976 ... and was the first practical method for establishing a shared secret over an unprotected communications channel." In 2002 [64], Martin Hellman gave credit to Merkle as well: "The system ... has since become known as Diffie-Hellman key exchange. While that system was first described in a paper by Diffie and me, it is a public key distribution system, a concept developed by Merkle, and hence should be called 'Diffie-Hellman-Merkle key exchange' if names are to be associated with it. I hope this small pulpit might help in that endeavor to recognize Merkle's equal contribution to the invention of public key cryptography."

U. S. Patent 4,200,770, now expired, describes the algorithm and credits Hellman, Diffie, and Merkle as inventors.

The simplest, and original, implementation of the protocol uses the multiplicative group of integers modulo $p$, where $p$ is prime and $g$ is primitive mod $p$. A more general description of the protocol uses an arbitrary finite cyclic group.

1. Alice and Bob agree on a finite cyclic group $G$ and a generating element $g$ in $G$. We will write the group $G$ multiplicatively.

2. Alice picks a random natural number $a$ and sends $g^a$ to Bob.

3. Bob picks a random natural number $b$ and sends $g^b$ to Alice.

4. Alice computes $K_A = (g^b)^a = g^{ba}$.

5. Bob computes $K_B = (g^a)^b = g^{ab}$.

Since $ab = ba$ (because $\mathbb{Z}$ is commutative), both Alice and Bob are now in possession of the same group element $K = K_A = K_B$ which can serve as the shared secret key.

The protocol is considered secure against eavesdroppers if $G$ and $g$ are chosen properly. The eavesdropper, Eve, must solve the *Diffie-Hellman problem* (recover

$g^{ab}$ from $g^a$ and $g^b$) to obtain the shared secret key. This is currently considered difficult for a "good" choice of parameters (see e.g., [100] for details).

An efficient algorithm to solve the *discrete logarithm problem* (i.e., recovering $a$ from $g$ and $g^a$) would obviously solve the Diffie-Hellman problem, making this and many other public key cryptosystems insecure. However, it is not known whether or not the discrete logarithm problem is *equivalent* to the Diffie-Hellman problem.

We note that there is a "brute force" method for solving the discrete logarithm problem: the eavesdropper Eve can just go over natural numbers $n$ from 1 up one at a time, compute $g^n$ and see whether she has a match with the transmitted element. This will require $O(|g|)$ multiplications, where $|g|$ is the order of $g$. Since in practical implementations $|g|$ is typically about $10^{300}$, this method is computationally infeasible.

This raises a question of computational efficiency for legitimate parties: on the surface, it looks like legitimate parties, too, have to perform $O(|g|)$ multiplications to compute $g^a$ or $g^b$. However, there is a faster way to compute $g^a$ for a particular $a$ by using the "square-and-multiply" algorithm, based on the binary form of $a$. For example, $g^{22} = (((g^2)^2)^2)^2 \cdot (g^2)^2 \cdot g^2$. Thus, to compute $g^a$, one actually needs $O(\log_2 a)$ multiplications, which is quite feasible.

## 1.3   The ElGamal cryptosystem

The ElGamal cryptosystem [35] is a public key cryptosystem which is based on the Diffie-Hellman key establishment (see the previous section). The ElGamal protocol is used in the free GNU Privacy Guard software, recent versions of PGP, and other cryptosystems. The Digital Signature Algorithm is a variant of the ElGamal signature scheme, which should not be confused with the ElGamal encryption protocol that we describe below.

1. Alice and Bob agree on a finite cyclic group $G$ and a generating element $g$ in $G$.

2. Alice (the receiver) picks a random natural number $a$ and publishes $c = g^a$.

3. Bob (the sender), who wants to send a message $m \in G$ (called a "plaintext" in cryptographic lingo) to Alice, picks a random natural number $b$ and sends two elements, $m \cdot c^b$ and $g^b$, to Alice. Note that $c^b = g^{ab}$.

4. Alice recovers $m = (m \cdot c^b) \cdot ((g^b)^a)^{-1}$.

A notable feature of the ElGamal encryption is that it is *probabilistic*, meaning that a single plaintext can be encrypted to many possible ciphertexts.

We also point out that the ElGamal encryption has an average expansion factor of 2. (Compare this to the encryption described in Section 1.1.)

## 1.4 Authentication

Authentication is the process of attempting to verify the digital identity of the sender of a communication. Of particular interest in public-key cryptography are *zero-knowledge* proofs of identity. This means that if the identity is true, no malicious verifier learns anything other than this fact. Thus, one party (the prover) wants to prove its identity to a second party (the verifier) via some secret information (a private key), but doesn't want anybody to learn anything about this secret.

Many key establishment protocols can be (slightly) modified to become authentication protocols. We illustrate this on the example of the Diffie-Hellman key establishment protocol (see Section 1.2).

Suppose Alice is the prover and Bob is the verifier, so that Alice wants to convince Bob that she knows a secret without revealing the secret itself.

1. Alice publishes a finite cyclic group $G$ and a generating element $g$ in $G$. Then she picks a random natural number $a$ and publishes $g^a$.

2. Bob picks a random natural number $b$ and sends a *challenge* $g^b$ to Alice.

3. Alice responds with a proof $P = (g^b)^a = g^{ba}$.

4. Bob verifies: $(g^a)^b = P$?.

We see that this protocol is almost identical to the Diffie-Hellman key establishment protocol. Later, in Chapter 4, we will see an example of an "independent" authentication protocol, which is not just a modification of a key establishment protocol.

# Chapter 2

# Background on Combinatorial Group Theory

In this chapter, we first give the definition of a free group, and then give a brief exposition of several classical techniques in combinatorial group theory, namely methods of Nielsen, Schreier, Whitehead, and Tietze. We do not go into details here because there are two very well-established monographs where a complete exposition of these techniques is given. For an exposition of Nielsen's and Schreier's methods, we recommend [96], whereas [95] has, in our opinion, a better exposition of Whitehead's and Tietze's methods.

Then, in Section 2.3, we describe algorithmic problems of group theory that will be exploited in Chapters 4 and 11 of this book.

In the concluding Section 2.6, we touch upon *normal forms* of group elements as a principal hiding mechanism for cryptographic protocols.

## 2.1 Basic definitions and notation

Let $G$ be a group. If $H$ is a subgroup of $G$, we write $H \leq G$; if $H$ is a normal subgroup of $G$, we write $H \trianglelefteq G$. For a subset $A \subseteq G$, by $\langle A \rangle$ we denote the subgroup of $G$ generated by $A$ (the intersection of all subgroups of $G$ containing $A$). It is easy to see that

$$\langle A \rangle = \{a_{i_1}^{\varepsilon_1}, \ldots, a_{i_n}^{\varepsilon_n} \mid a_{i_j} \in A, \ \varepsilon_j \in \{1, -1\}, \ n \in \mathbb{N}\}.$$

Let $X$ be an arbitrary set. A *word* $w$ in $X$ is a finite (possibly empty) sequence of elements that we write as $w = y_1 \ldots y_n$, $y_i \in X$. The number $n$ is called the *length* of the word $w$; we denote it by $|w|$. We denote the empty word by $\varepsilon$ and put $|\varepsilon| = 0$. Then, let $X^{-1} = \{x^{-1} \mid x \in X\}$, where $x^{-1}$ is just a formal expression obtained from $x$ and $-1$. If $x \in X$, then the symbols $x$ and $x^{-1}$ are called *literals* in $X$. Denote by $X^{\pm 1} = X \cup X^{-1}$ the set of all literals in $X$.

An expression of the form

$$w = x_{i_1}^{\varepsilon_1} \cdots x_{i_n}^{\varepsilon_n}, \tag{2.1}$$

where $x_{i_j} \in X, \varepsilon_j \in \{1, -1\}$ is called a *group word* in $X$. So a group word in $X$ is just a word in the alphabet $X^{\pm 1}$.

A group word $w = y_1 \cdots y_n$ is *reduced* if for any $i = i, \ldots, n-1$, $y_i \neq y_{i+1}^{-1}$, that is, $w$ does not contain a subword of the form $yy^{-1}$ for any literal $y \in X^{\pm 1}$. We assume that the empty word is reduced.

If $X \subseteq G$, then every group word $w = x_{i_1}^{\varepsilon_1} \cdots x_{i_n}^{\varepsilon_n}$ in $X$ determines a unique element of $G$ which is equal to the product $x_{i_1}^{\varepsilon_1} \cdots x_{i_n}^{\varepsilon_n}$ of the elements $x_{i_j}^{\varepsilon_j} \in G$. By convention, the empty word $\varepsilon$ determines the identity 1 of $G$.

**Definition 2.1.1.** A group $G$ is called a *free group* if there is a generating set $X$ of $G$ such that every nonempty reduced group word in $X$ defines a nontrivial element of $G$.

In this case $X$ is called a *free basis* of $G$ and $G$ is called *a free group on $X$*, or *a group freely generated by $X$*. It follows from the definition that every element of $F(X)$ can be defined by a reduced group word in $X$. Moreover, different reduced words in $X$ define different elements of $G$.

Free groups have the following *universal property:*

**Theorem 2.1.2.** *Let $G$ be a group with a generating set $X \subseteq G$. Then $G$ is free on $X$ if and only if the following universal property holds: every map $\varphi : X \to H$ from $X$ into a group $H$ can be extended to a unique homomorphism*

$$\varphi^* : G \to H,$$

*so that the diagram below is commutative:*

*(here $X \xrightarrow{i} G$ is the natural inclusion of $X$ into $G$).*

**Corollary 2.1.3.** *Let $G$ be a free group on $X$. Then the identity map $X \to X$ extends to an isomorphism $G \to F(X)$.*

This corollary allows us to identify a free group freely generated by $X$ with the group $F(X)$ of reduced group words in $X$. In what follows we usually call the group $F(X)$ a free group on $X$.

## 2.2 Presentations of groups by generators and relators

The universal property of free groups allows one to describe arbitrary groups in terms of *generators* and *relators*.

Let $G$ be a group with a generating set $X$. By the universal property of free groups there exists a homomorphism $\psi : F(X) \to G$ such that $\psi(x) = x$ for $x \in X$. It follows that $\psi$ is onto, so by the first isomorphism theorem

$$G \simeq F(X)\big/\mathrm{ker}(\psi).$$

In this case $\mathrm{ker}(\psi)$ is viewed as the set of relators of $G$, and a group word $w \in \mathrm{ker}(\psi)$ is called a *relator* of $G$ in generators $X$. If a subset $R \subseteq \mathrm{ker}(\psi)$ generates $\mathrm{ker}(\psi)$ as a normal subgroup of $F(X)$ then it is termed a set of *defining relators* of $G$ relative to $X$. The pair $\langle X \mid R \rangle$ is called a *presentation* of a group $G$; it determines $G$ uniquely up to isomorphism. The presentation $\langle X \mid R \rangle$ is finite if both sets $X$ and $R$ are finite. A group is *finitely presented* if it has at least one finite presentation. Presentations provide a universal method to describe groups. In particular, finitely presented groups admit finite descriptions, e. g.

$$G = \langle x_1, x_2, \ldots, x_n \mid r_1, r_2, \ldots, r_k \rangle.$$

All finitely generated abelian groups are finitely presented (a group $G$ is *abelian*, or commutative, if $ab = ba$ for all $a, b \in G$). Other examples of finitely presented groups include finitely generated nilpotent groups (see our Section 6.1.6), braid groups (Section 5.1), Thompson's group (Section 5.2).

## 2.3 Algorithmic problems of group theory

Algorithmic problems of (semi)group theory that we consider in this section are of two different kinds:

1. *Decision problems* are problems of the following nature: given a property $\mathcal{P}$ and an object $\mathcal{O}$, find out whether or not the object $\mathcal{O}$ has the property $\mathcal{P}$.

2. *Search problems* are of the following nature: given a property $\mathcal{P}$ and the information that there are objects with the property $\mathcal{P}$, find at least one particular object with the property $\mathcal{P}$.

We are now going to discuss several particular algorithmic problems of group theory that have been used in cryptography.

### 2.3.1 The word problem

The *word problem* (WP) is: given a recursive presentation of a group $G$ and an element $g \in G$, find out whether or not $g = 1$ in $G$.

From the very description of the word problem we see that it consists of two parts: "whether" and "not". We call them the "yes" and "no" parts of the word problem, respectively. If a group is given by a recursive presentation in terms of generators and relators, then the "yes" part of the word problem has a recursive solution:

**Proposition 2.3.1.** *Let $\langle X; R \rangle$ be a recursive presentation of a group $G$. Then the set of all words $g \in G$ such that $g = 1$ in $G$ is recursively enumerable.*

The *word search problem* (WSP) is: given a recursive presentation of a group $G$ and an element $g = 1$ in $G$, find a presentation of $g$ as a product of conjugates of defining relators and their inverses.

We note that the word search problem always has a recursive solution because one can recursively enumerate all products of defining relators, their inverses and conjugates. However, the number of factors in such a product required to represent a word of length $n$ which is equal to 1 in $G$, can be very large compared to $n$; in particular, there are groups $G$ with efficiently solvable word problem and words $w$ of length $n$ equal to 1 in $G$, such that the number of factors in any factorization of $w$ into a product of defining relators, their inverses and conjugates is not bounded by any tower of exponents in $n$, see [117]. Furthermore, if in a group $G$ the word problem is recursively unsolvable, then the length of a proof verifying that $w = 1$ in $G$ is not bounded by any recursive function of the length of $w$.

## 2.3.2   The conjugacy problem

The next two problems of interest to us are

The *conjugacy problem* (CP) is: given a recursive presentation of a group $G$ and two elements $g, h \in G$, find out whether or not there is an element $x \in G$ such that $x^{-1}gx = h$.

Again, just as the word problem, the conjugacy problem consists of the "yes" and "no" parts, with the "yes" part always recursive because one can recursively enumerate all conjugates of a given element.

The *conjugacy search problem* (CSP) is: given a recursive presentation of a group $G$ and two conjugate elements $g, h \in G$, find a particular element $x \in G$ such that $x^{-1}gx = h$.

As we have already mentioned, the conjugacy search problem always has a recursive solution because one can recursively enumerate all conjugates of a given element, but as with the word search problem, this kind of solution can be extremely inefficient.

## 2.3.3   The decomposition and factorization problems

One of the natural ramifications of the conjugacy search problem is the following:

The *decomposition search problem*: given a recursive presentation of a group $G$, two recursively generated subgroups $A, B \leq G$, and two elements $g, h \in G$, find two elements $x \in A$ and $y \in B$ that would satisfy $x \cdot g \cdot y = h$, provided at least one such pair of elements exists.

We note that *some* $x$ and $y$ satisfying the equality $x \cdot g \cdot y = h$ always exist (e. g. $x = 1$, $y = g^{-1}h$), so the point is to have them satisfy the conditions $x \in A$, $y \in B$. We therefore will not usually refer to this problem as a *subgroup-restricted* decomposition search problem because it is always going to be subgroup-restricted; otherwise it does not make much sense.

A special case of the decomposition search problem, where $A = B$, is also known as the *double coset problem*.

Clearly, the decomposition problem also has the decision version. So far, it has not been used in cryptography.

One more special case of the decomposition search problem, where $g = 1$, deserves special attention.

The *factorization problem*: given an element $w$ of a recursively presented group $G$ and two subgroups $A, B \leq G$, find out whether or not there are two elements $a \in A$ and $b \in B$ such that $a \cdot b = w$.

The *factorization search problem*: given an element $w$ of a recursively presented group $G$ and two recursively generated subgroups $A, B \leq G$, find any two elements $a \in A$ and $b \in B$ that would satisfy $a \cdot b = w$, provided at least one such pair of elements exists.

There are relations between the algorithmic problems discussed so far, which we summarize in the following

**Proposition 2.3.2.** *Let $G$ be a recursively presented group.*

1. *If the conjugacy problem in $G$ is solvable, then the word problem is solvable, too.*

2. *If the conjugacy search problem in $G$ is solvable, then the decomposition search problem is solvable for commuting subgroups $A, B \leq G$ (i.e., $ab = ba$ for all $a \in A, b \in B$).*

3. *If the conjugacy search problem in $G$ is solvable, then the factorization search problem is solvable for commuting subgroups $A, B \leq G$.*

The first statement of this proposition is obvious since conjugacy to the identity element 1 is the same as equality to 1. Two other statements are not immediately obvious; proofs are given in our Section 4.7.

## 2.3.4 The membership problem

Now we are getting to the next pair of problems.

The *membership problem*: given a recursively presented group $G$, a subgroup $H \leq G$ generated by $h_1, \ldots, h_k$, and an element $g \in G$, find out whether or not $g \in H$.

Again, the membership problem consists of the "yes" and "no" parts, with the "yes" part always recursive because one can recursively enumerate all elements of a subgroup given by finitely many generators.

We note that the membership problem also has a less descriptive name, "the generalized word problem".

The *membership search problem*: given a recursively presented group $G$, a subgroup $H \leq G$ generated by $h_1, \ldots, h_k$, and an element $h \in H$, find an expression of $h$ in terms of $h_1, \ldots, h_k$.

In the next Section 2.4, we are going to show how the membership problem can be solved for any finitely generated subgroup of any free group.

### 2.3.5   The isomorphism problem

Finally, we mention the isomorphism problem that will be important in our Chapter 11.

The *isomorphism problem* is: given two finitely presented groups $G_1$ and $G_2$, find out whether or not they are isomorphic.

We note that Tietze's method described in our Section 2.5 provides a recursive enumeration of all finitely presented groups isomorphic to a given finitely presented group, which implies that the "yes" part of the isomorphism problem is always recursive.

On the other hand, specific instances of the isomorphism problem *may* provide examples of group-theoretic decision problems, both the "yes" and "no" parts of which are nonrecursive. Here we can offer a candidate problem of that kind:

**Problem 2.3.3.** [5, Problem (A5)] Is a given finitely presented group metabelian?

Metabelian groups and their properties are discussed in our Section 5.5.

## 2.4   Nielsen's and Schreier's methods

Let $F = F_n$ be the free group of a finite rank $n \geq 2$ with a set $X = \{x_1, \ldots, x_n\}$ of free generators. Let $Y = \{y_1, \ldots, y_m\}$ be an arbitrary finite set of elements of the group $F$. Consider the following elementary transformations that can be applied to $Y$:

**(N1)** $y_i$ is replaced by $y_i y_j$ or by $y_j y_i$ for some $j \neq i$;

**(N2)** $y_i$ is replaced by $y_i^{-1}$;

**(N3)** $y_i$ is replaced by some $y_j$, and at the same time $y_j$ is replaced by $y_i$;

**(N4)** delete some $y_i$ if $y_i = 1$.

It is understood that $y_j$ does not change if $j \neq i$.

Every finite set of reduced words of $F_n$ can be carried by a finite sequence of Nielsen transformations to a *Nielsen-reduced* set $U = \{u_1, \ldots, u_k\}$; i.e., a set such that for any triple $v_1, v_2, v_3$ of the form $u_i^{\pm 1}$, the following three conditions hold:

(i) $v_1 \neq 1$;

(ii) $v_1 v_2 \neq 1$ implies $|v_1 v_2| \geq |v_1|, |v_2|$;

(iii) $v_1 v_2 \neq 1$ and $v_2 v_3 \neq 1$ implies $|v_1 v_2 v_3| > |v_1| - |v_2| + |v_3|$.

It is easy to see that if $U = (u_1, u_2, \ldots, u_k)$ is Nielsen-reduced, then the subgroup of $F_n$ generated by $U$ is free with a basis $U$, see e.g. [96].

One might notice that some of the transformations (N1)–(N4) are redundant; i.e., they are compositions of other ones. The reason behind that will be explained below.

We say that two sets $Y$ and $\tilde{Y}$ are Nielsen equivalent if one of them can be obtained from another by applying a sequence of transformations (N1)–(N4). It was proved by Nielsen that two sets $Y$ and $\tilde{Y}$ generate the same subgroup of the group $F$ if and only if they are Nielsen equivalent. This result is now one of the central points in combinatorial group theory.

Note, however, that this result alone does not give an *algorithm* for deciding whether or not $Y$ and $\tilde{Y}$ generate the same subgroup of $F$. To obtain an algorithm, we need to somehow define the *complexity* of a given set of elements and then show that a sequence of Nielsen transformations (N1)–(N4) can be arranged so that this complexity decreases (or, at least, does not increase) *at every step* (this is where we may need "redundant" elementary transformations!).

This was also done by Nielsen; the complexity of a given set $Y = \{y_1, \ldots, y_m\}$ is just the sum of the lengths of the words $y_1, \ldots, y_m$. Now the algorithm is as follows. Reduce both sets $Y$ and $\tilde{Y}$ to Nielsen-reduced sets. This procedure is finite because the sum of the lengths of the words decreases at every step. Then solve the membership problem for every element of $Y$ in the subgroup generated by $\tilde{Y}$ and vice versa.

Nielsen's method therefore yields (in particular) an algorithm for deciding whether or not a given endomorphism of a free group of finite rank is an automorphism.

Another classical method that we sketch in this section is that of Schreier. We are going to give a brief exposition of this method here in a special case of a free group; however, it is valid for arbitrary groups as well.

Let $H$ be a subgroup of $F$. A *right coset representative function* for $F$ (on the generators $x_i$) modulo $H$ is a mapping of words in $x_i$, $w(x_1, x_2, \ldots) \rightarrow \overline{w}(x_1, x_2, \ldots)$, where the $\overline{w}(x_1, x_2, \ldots)$ form a right coset representative system for $F$ modulo $H$, which contains the empty word, and where $\overline{w}(x_1, x_2, \ldots)$ is the representative of the coset of $w(x_1, x_2, \ldots)$. Then we have:

**Theorem 2.4.1.** *If $w \to \overline{w}$ is a right coset function for $F$ modulo $H$, then $H$ is generated by the words*

$$ux_i \cdot \overline{ux_i}^{-1},$$

*where $u$ is an arbitrary representative and $x_i$ is a generator of $F$.*

This already implies, for instance, that if $F$ is finitely generated and $H$ is a subgroup of finite index, then $H$ is finitely generated.

Furthermore, a *Schreier right coset function* is one for which any initial segment of a representative is again a representative. The system of representatives is then called a *Schreier system*. It can be shown that there is always some Schreier system of representatives for $F$ modulo $H$. Also, there is a *minimal Schreier system*; i.e., a Schreier system in which each representative has a length not exceeding the length of any word it represents. Not every Schreier system is minimal.

**Example 2.4.2.** Let $F$ be the free group on $a$ and $b$, and let $H$ be the normal subgroup of $F$ generated by $a^2$, $b^2$, and $aba^{-1}b^{-1}$. Then $F/H$ has four cosets. The representative system $\{1, a, b, ab\}$ is Schreier, as is the representative system $\{1, a, b, ab^{-1}\}$. The representative system $\{1, a, b, a^{-1}b^{-1}\}$ is not Schreier; for, the initial segment $a^{-1}$ of $a^{-1}b^{-1}$ is not a representative.

Using a *Reidemeister–Schreier rewriting process*, one can obtain a presentation (by generators and relators) for $H$. This implies, among other things, the following important result:

**Theorem 2.4.3 (Nielsen–Schreier).** *Every nontrivial subgroup $H$ of a free group $F$ is free.*

One can effectively obtain a set of free generating elements for $H$; namely, those $ux_i \cdot \overline{ux_i}^{-1}$ such that $ux_i$ is not freely equal to a representative. Schreier obtained this set of free generators for $H$ in 1927. In 1921, Nielsen, using quite a different method, had constructed a set of free generators for $H$ if $F$ was finitely generated. The generators of Nielsen are precisely those Schreier generators obtained when a minimal Schreier system is used.

We refer to [96, Chapter 3] for more details.

## 2.5   Tietze's method

Attempting to solve one of the major problems of combinatorial group theory, the isomorphism problem, Tietze introduced isomorphism-preserving elementary transformations that can be applied to groups presented by generators and relators.

Let

$$G = \langle x_1, x_2, \dots \mid r_1, r_2, \dots \rangle$$

be a presentation of a group $G = F/R$, where $F$ is the ambient free group generated by $x_1, x_2, \dots$, and $R$ is the normal closure of $r_1, r_2, \dots$; i.e., the smallest normal subgroup of $F$ containing $r_1, r_2, \dots$.

The elementary transformations are of the following types.

**(T1)** *Introducing a new generator*: Replace $\langle x_1, x_2, \ldots \mid r_1, r_2, \ldots \rangle$ by $\langle y, x_1, x_2, \ldots \mid ys^{-1}, r_1, r_2, \ldots \rangle$, where $s = s(x_1, x_2, \ldots)$ is an arbitrary element in the generators $x_1, x_2, \ldots$.

**(T2)** *Canceling a generator* (this is the converse of (T1)): If we have a presentation of the form $\langle y, x_1, x_2, \ldots \mid q, r_1, r_2, \ldots \rangle$, where $q$ is of the form $ys^{-1}$, and $s, r_1, r_2, \ldots$ are in the group generated by $x_1, x_2, \ldots$, replace this presentation by $\langle x_1, x_2, \ldots \mid r_1, r_2, \ldots \rangle$.

**(T3)** *Applying an automorphism*: Apply an automorphism of the free group generated by $x_1, x_2, \ldots$ to all the relators $r_1, r_2, \ldots$.

**(T4)** *Changing defining relators*: Replace the set $r_1, r_2, \ldots$ of defining relators by another set $r_1', r_2', \ldots$ with the same normal closure. That means, each of $r_1', r_2', \ldots$ should belong to the normal subgroup generated by $r_1, r_2, \ldots$, and vice versa.

Then we have the following useful result due to Tietze (see, e.g., [95]):

**Theorem 2.5.1.** *Two groups $\langle x_1, x_2, \ldots \mid r_1, r_2, \ldots \rangle$ and $\langle x_1, x_2, \ldots \mid s_1, s_2, \ldots \rangle$ are isomorphic if and only if one can get from one of the presentations to the other by a sequence of transformations* (T1)–(T4).

For most practical purposes, groups under consideration are finitely presented, in which case there exists a finite sequence of transformations (T1)–(T4) taking one of the presentations to the other. Still, Theorem 2.5.1 does not give any constructive procedure for deciding in a finite number of steps whether one finite presentation can be obtained from another by Tietze transformations because, for example, there is no indication of how to select the element $S$ in a transformation (T1). Thus Theorem 2.5.1 does not yield a solution to the isomorphism problem.

However, it has been used in many situations to derive various invariants of isomorphic groups, most notably Alexander polynomials that turned out to be quite useful in knot theory.

Also, Tietze's method gives an easy, practical way of constructing "exotic" examples of isomorphic groups that helped to refute several conjectures in combinatorial group theory. For a similar reason, Tietze transformations can be useful for "diffusing" presentations of groups in some cryptographic protocols (see our Section 6.1.3).

## 2.6 Normal forms

Normal forms of group elements are principal hiding mechanisms for cryptographic protocols.

A normal form is required to have two essential properties: (1) every object under consideration must have exactly one normal form, and (2) two objects that have the same normal form must be the same up to some equivalence. The uniqueness requirement in (1) is sometimes relaxed, allowing the normal form to be unique up to some simple equivalence.

Normal forms may be "natural" and simple, but they may also be quite elaborate. We give some examples of normal forms below.

**Example 2.6.1.** In the (additive) group of integers, we have many "natural" normal forms: decimal, binary, etc. These are quite good for hiding factors in a product; for example, in the product $3 \cdot 7 = 21$ we do not see 3 or 7. We make one more observation here which is important from the point of view of cryptography: if there are several different normal forms for elements of a given group, then one normal form might reveal what another one is trying to conceal. For example, the number 31 in the decimal form "looks random", but the same number in the binary form has a clear pattern: 11111. We re-iterate this point in Sections 5.1 and 5.2 of this book, in more complex situations.

There are also other normal forms for integers; for example, every integer is a product of primes. This factorization is not unique, but if we require, in addition, that a product of primes should be *in increasing order*, then it becomes a unique normal form.

**Example 2.6.2.** In a group of matrices over a ring $R$, every matrix is the normal form for itself. This normal form is unique up to the equality of the entries in the ring $R$.

**Example 2.6.3.** In some groups given by generators and defining relators (cf. our Section 2.2), there are *rewriting systems*, i.e., procedures which take a word in a given alphabet as input and transform it to another word in the same alphabet by using defining relators. This procedure terminates with the normal form of the group element represented by the input word. We give an example of a rewriting system in Thompson's group in Section 5.2 of this book.

In other groups given by generators and defining relators, normal forms may be based on some special (topological, or geometric, or other) properties of a given group, and not just on a rewriting system. A good example of this sort is provided by braid groups, see our Section 5.1. There are several different normal forms for elements of a braid group; the classical one, called the Garside normal form, is not even a word in generators of the group. We cannot give more details here without introducing a large amount of background material, so we just refer the interested reader to the monographs [10] and [30].

# Chapter 3

# Background on Computational Complexity

## 3.1 Algorithms

In all instances, if not said otherwise, we use Turing machines as our principal model of computation. In this section we briefly recall some basic definitions and notation concerning Turing machines that are used throughout this chapter.

### 3.1.1 Deterministic Turing machines

In this section we give basic definitions of deterministic and non-deterministic Turing machines to be used in the sequel.

**Definition 3.1.1.** A one-tape *Turing machine* (TM) $M$ is a 5-tuple $\langle Q, \Sigma, s, f, \delta \rangle$ where:

- $Q$ is a finite set of *states*;

- $\Sigma$ is a finite set of the *tape alphabet*;

- $s \in Q$ is the *initial state*;

- $f \in Q$ is the *final state*;

- $\delta : Q \times \Sigma \to Q \times \Sigma \times \{L, R\}$ called the *transition function*.

Additionally, $M$ uses a blank symbol $\sqcup$ different from the symbols in $\Sigma$ to mark the parts of the infinite tape that are not in use.

We can define the operation of a TM formally using the notion of a configuration which contains a complete description of the current state of computation. A *configuration* of $M$ is a triple $(q, w, u)$, where $w, u$ are $\Sigma$-strings and $q \in Q$.

- $w$ is a string to the left of the head;

- $u$ is the string to the right of the head, including the symbol scanned by the head;

- $q$ is the current state.

We say that a configuration $(q, w, u)$ *yields* a configuration $(q', w', u')$ in one step, denoted by $(q, w, u) \xrightarrow{M} (q', w', u')$, if a step of the machine from the configuration $(q, w, u)$ results in the configuration $(q', w', u')$. Using the relation "yields in one step" we can define relations "yields in $k$ steps", denoted by $\xrightarrow{M^k}$, and "yields", denoted by $\xrightarrow{M^*}$. A sequence of configurations that $M$ yields on the input $x$ is called an *execution flow* of $M$ on $x$.

We say that $M$ *halts* on $x \in \Sigma^*$ if the configuration $(s, \varepsilon, x)$ yields a configuration $(f, w, u)$ for some $\Sigma$-strings $w$ and $u$; in this case, $(f, w, u)$ is called a *halting configuration*. The number of steps $M$ performs on a $\Sigma$-string $x$ before it stops is denoted by $T_M(x)$. The *halting problem* for $M$ is an algorithmic problem to determine whether $M$ halts or not, i.e., whether $T_M(x) = \infty$ or not.

We say that a TM $M$ *solves* or *decides* a decision problem $D$ over an alphabet $\Sigma$ if $M$ stops on every input $x \in \Sigma^*$ with an answer

- *Yes* (i.e., at configuration $(f, \varepsilon, 1)$) if $x$ is a positive instance of $D$;

- *No* (i.e., at configuration $(f, \varepsilon, 0)$) otherwise.

We say that $M$ *partially decides* $D$ if it decides $D$ correctly on a subset $D'$ of $D$ and on $D - D'$ it either does not stop or stops with an answer *DontKnow* (i.e., stops at configuration $(f, \varepsilon, \sqcup)$).

## 3.1.2 Non-deterministic Turing machines

**Definition 3.1.2.** A one-tape *non-deterministic Turing machine* (NTM) $M$ is a 5-tuple $\langle Q, \Sigma, s, f, \delta \rangle$, where $Q, \Sigma, s, f$ and $\delta$ are as in the definition of deterministic TM except that $\delta : Q \times \Gamma \to Q \times \Gamma \times \{L, R\}$ is a multivalued function (or a binary relation).

Configurations of an NTM are the same as configurations of a deterministic TM, but the relation "yields in one step" is slightly different. We say that an NTM $M$, given a configuration $c_1 = (q, w, u)$, yields a configuration $c_2 = (q', w', u')$ if there exists a transition rule which transforms $c_1$ to $c_2$. As before define the relation "yields" based on the relation "yields in one step". Note that, according to this definition of a configuration, an NTM can yield two or more configurations in one step and exponentially many configurations in $n$ steps. Moreover, it is allowed to yield both halting and non-halting configurations on the same input, which means we need a revised definition of acceptance. Thus, we will say that $M$ *accepts* an input $x$ if the initial configuration yields a halting configuration.

### 3.1.3 Probabilistic Turing machines

Intuitively, a *probabilistic Turing machine* is a Turing machine with a random number generator. More precisely, it is a machine with two transition functions $\delta_1$ and $\delta_2$. At each step of computation, each function $\delta_i$ is used with probability $\frac{1}{2}$. Probabilistic Turing machines are similar to nondeterministic Turing machines in the sense that one configuration can yield many configurations in $n$ steps, with the difference of how we interpret computations. For an NTM $M$ we are interested in the question whether or not there is a sequence of choices that make $M$ accept a certain input, whereas for the same PTM $M$ the question is with what probability acceptance occurs.

The output of a probabilistic machine $M$ on input $x$ is a random variable, denoted by $M(x)$. By $\mathbf{P}(M(x) = y)$ we denote the probability for the machine $M$ to output $y$ on the input $x$. The probability space is the space of all possible outcomes of the internal coin flips of $M$ taken with uniform probability distribution.

**Definition 3.1.3.** Let $D$ be a decision problem. We say that a PTM $M$ *decides* $D$ if it outputs the right answer with probability at least $2/3$.

## 3.2 Computational problems

In this section we briefly discuss general definitions of computational (or algorithmic) problems, size functions and stratification. At the end of the section we recall basics of the worst case analysis of algorithms, introduce the worst case complexity, and discuss its limitations. One of the main purposes of this section is to introduce notation and terminology.

### 3.2.1 Decision and search computational problems

We start with the standard definitions of decision and search computational problems, though presented in a slightly more general, than usual, form.

Let $X$ be a finite alphabet and $X^*$ the set of all words in the alphabet $X$. Sometimes subsets of $X^*$ are called *languages* in $X$. A *decision problem* for a subset $L \subseteq X^*$ is the following problem:

> Is there an algorithm that for a given word $w \in X^*$ determines whether $w$ belongs to $L$ or not?

If such an algorithm exists, we call it a *decision algorithm* for $L$, and in this case the decision problem for $L$, as well as the language $L$, is called *decidable*. More formally, a decision problem is given by a pair $\mathcal{D} = (L, X^*)$, with $L \subseteq X^*$. We refer to words from $X^*$ as *instances* of the decision problem $\mathcal{D}$. The set $L$ is termed the *positive*, or the *Yes*, part of the problem $\mathcal{D}$, and the complement $\bar{L}(\mathcal{D}) = X^* - L$ is the *negative*, or the *No*, part of $\mathcal{D}$. If there are no decision algorithms for $\mathcal{D}$, then $\mathcal{D}$ is called *undecidable*.

In practice, many decision problems appear in the *relativized* form $\mathcal{D} = (L, U)$, where $L \subseteq U \subseteq X^*$, so the set of instances of $\mathcal{D}$ is restricted to the subset $U$. For example, the famous Diophantine Problem for integers (10th Hilbert's problem) asks whether a given polynomial with integer coefficients has an integer root or not. In this case polynomials are usually presented by specific words (terms) in the corresponding alphabet and there is no need to consider arbitrary words as instances. Typically, the set $U$ is assumed to be decidable, or at least effectively enumerable.

Sometimes decision problems occur in a more general form given by a pair $\mathcal{D} = (L, U)$, where $L \subseteq U$ and $U$ is a subset of the Cartesian product $X^* \times \cdots \times X^*$ of $k \geq 1$ copies of $X^*$. For example, the conjugacy problem in a given group is naturally described by a set of pairs of elements of the group; or the isomorphism problem for graphs is usually given by a set of pairs of graphs, etc. By introducing an extra alphabet symbol "," one could view a $k$-tuple of words $(w_1, w_2, \ldots, w_k) \in (X^*)^k$ as a single word in the alphabet $X' = X \cup \{,\}$, which allows one to view the decision problems above as given, again, in the form $(L, U)$, with $U \subseteq (X')^*$.

A *search computational problem* can be described by a binary predicate $R(x, y) \subseteq X^* \times Y^*$, where $X$ and $Y$ are finite alphabets. In this case, given an input $x \in X^*$, one has to find a word $y \in Y^*$ such that $R(x, y)$ holds. For example, in the Diophantine Problem above $x$ is a polynomial equation (or a "word" describing this equation) $E_x(y) = 0$ in a tuple of variables $y$, and the predicate $R(x, y)$ holds for given values of $x$ and $y$ if and only if these values give a solution of $E_x(y) = 0$. In what follows we always assume that the predicate $R(x, y)$ is computable.

In general, one can consider two different variations of search problems. The first one requires, for a given $x \in X^*$, to decide first whether there exists $y \in Y^*$ such that $R(x, y)$ holds, and only after that to find such $y$ if it exists. In this case the search problem $\mathcal{D} = (R, X^* \times Y^*)$ contains the decision problem $\exists y R(x, y)$ as a subproblem. In the second variation one assumes from the beginning that for a given $x$ the corresponding $y$ always exists and the problem is just to find such $y$. The latter can be described in the relativized form by $\mathcal{D} = (R, U \times Y^*)$, where $U$ is the "Yes" part of the decision problem $\exists y R(x, y)$.

Quite often algorithmic search problems occur as "decision problems with witnesses". For example, given a pair of elements $(x_1, x_2)$ of a given groups $G$, one has to check if they are conjugate in $G$ or not, and if they are, one has to find a "witness" to that, i.e., one of the conjugating elements. Similarly, given two finite graphs $\Gamma_1$ and $\Gamma_2$ one has to solve the isomorphism problem for $\Gamma_1, \Gamma_2$, and if the graphs are isomorphic, to find an isomorphism witnessing the solution.

If the predicate $R(x, y)$ is computable, then the second variation of the corresponding search problem is always decidable (it suffices to verify all possible inputs from $Y^*$ one by one to find a solution if it exists). Note also that in this case the first variation is also decidable provided the decision problem $\exists y R(x, y)$ is decidable. Hence, in this situation the decidability of the search algorithmic problems follows from the decidability of the corresponding decision problems. However, in many cases our main concern is about complexity of the decision or search algorithms, so in this respect, search computational problems a priori cannot be easily reduced to decision problems.

Our final remark on descriptions of computational problems is that quite often inputs for a particular problem are not given by words in an alphabet; moreover, any representation of inputs by words, although possible, brings unnecessary complications into the picture. Furthermore, such a representation may even change the nature of the problem itself. For example, in computational problems for graphs it is possible to represent graphs by words in some alphabet, but it is not very convenient, and sometimes misleading (see [59] for details). In these cases we feel free to represent instances of the problem in a natural way, not encoding them by words. One can easily adjust all the definitions above to such representations. In the most general way, we view a computational decision problem as a pair $\mathcal{D} = (L, I)$, where $I = I_{\mathcal{D}}$ is a set of inputs for $\mathcal{D}$ and $L = L_{\mathcal{D}}$ is a subset of $I$, the "Yes" part of $\mathcal{D}$. Similarly, a general search computational problem can be described as $\mathcal{D} = (R(x, y), I)$, where $I$ is a subset of $I_1 \times I_2$ and $R(x, y) \subseteq I$. We always assume that the set $I$, as well as all its elements, allows an effective description. We do not want to go into details on this subject, but in every particular case this will be clear from the context. For example, we may discuss algorithms over matrices or polynomials with coefficients from finite fields or rational numbers, or finite graphs, or finitely presented groups, etc., without specifying any particular representations of these objects by words in a particular alphabet.

## 3.2.2 Size functions

In this section we discuss various ways to introduce the size, or complexity, of instances of algorithmic problems. This is part of a much bigger topic on complexity of descriptions of mathematical objects.

To study computational complexity of algorithms one has to be able to compare the resources $r_{\mathcal{A}}(x)$ spent by an algorithm $\mathcal{A}$ on a given input $x$ with the "size" $size(x)$ (or "complexity") of the input. In what follows we mostly consider only one resource: the time spent by the algorithm on a given input. To get rid of inessential details coming into the picture from a particular model of computation or the way the inputs are encoded, one can consider the growth of the "resource" function $r_{\mathcal{A}} : size(x) \to r(x)$. This is where the notion of the size of inputs plays an important role and greatly affects behavior of the function $r_{\mathcal{A}}$. Usually the size of an input $x$ depends on the way the inputs are described. For example, if a natural number $x$ is given in a unary number system, i.e., $x$ is viewed as a word of length $x$ in a unary alphabet $\{1\}$, say $x = 11 \ldots 1$, then it is natural to assume that the size of $x$ is the length of the word representing $x$, i.e., is $x$ itself. However, if the natural number $x$ is given, say, in the binary or decimal representation, then the size of $x$ (which is the length of the corresponding representation) is exponentially smaller (about $\log x$), so the same algorithm $\mathcal{A}$ may have quite different resource functions depending on the choice of the size function.

The choice of the size function depends, of course, on the problem $\mathcal{D}$. There are two principle approaches to define size functions on a set $I$. In the first one, the size of an input $x \in I$ is the *descriptive complexity* $d(x)$ of $x$, which is the length of the minimal description of $x$ among all natural representations of $x$ of a fixed type. For example, the length $|w|$ of a word $w$ in a given alphabet is usually viewed as the size of $w$. However, there are other natural size functions, for instance, the size of an element $g$ of a finitely generated group $G$ could be the length of a shortest product of elements and their inverses (from a fixed finite generating set of $G$) which is equal to $g$. One of the most intriguing size functions in this class comes from the so-called Kolmogorov complexity (see [94] for details). In the second approach, the size of an element $x \in I$ is taken to be the time required for a given generating procedure to generate $x$ (*production complexity*). For example, when producing keys for a cryptoscheme, it is natural to view the time spent on generating a key $x$ as the size $c(x)$ of $x$. In this case, the computational security of the scheme may be based on the amount of recourses required to break $x$ relative to the size $c(x)$. It could easily be that the descriptive complexity $d(x)$ of $x$ is small but the computational complexity $c(x)$ (with respect to a given generating procedure) is large.

In general, a *size* (or *complexity*) *function* on a set $I$ is an arbitrary nonnegative integral (or real) function $s : I \longrightarrow \mathbb{N}^+$ (or $s : I \longrightarrow \mathbb{R}^+$) that satisfies the following conditions:

C1) for every $n \in \mathbb{N}$ the preimage $s^{-1}(n)$ is either a finite subset of $I$ or, in the case where $I$ is equipped with a measure, a measurable subset of $I$. If $s$ is a size function with values in $\mathbb{R}^+$ and $I$ is equipped with a measure $\mu$, then we require (if not said otherwise) that $s$ is $\mu$-measurable.

C2) for an element $x \in I$ given in a fixed representation, one can effectively compute $s(x)$.

Note that it makes sense to only consider size functions on $I$ which are "natural" in the context of a problem at hand. This is an important principle that we do not formally include in the conditions above because of the difficulties with formalization. For an algorithmic problem $\mathcal{D}$, by $s_{\mathcal{D}}$ we denote the given size function on the set of instances $I$ (if it exists).

There is another very interesting way to define size functions on sets of inputs of algorithms, which is not yet developed enough. We briefly sketch the main idea here.

Let $\mathcal{C}$ be a class of decision or search algorithms solving a computational problem $\mathcal{D} = (L, I)$ that use different strategies but employ more or less similar tools. Let $A_{opt}$ be a non-deterministic algorithm, which is based on the same tools as algorithms from $\mathcal{C}$ are, but which is using an oracle that provides an optimal strategy at each step of computation. One can define the size $s_{opt}(x)$ of $x \in I$ as the time spent by the algorithm $A_{opt}$ on the input $x$. Then for any particular algorithm $A \in \mathcal{C}$ the resource function $r_A$ will allow one to evaluate the performance of the algorithm $A$ in comparison with the optimal (in the class $C$) "ideal" algorithm $A_{opt}$. This approach proved to be very useful, for example, in the study of algorithmic complexity of the famous Whitehead problem from group theory and topology (see [109]).

### 3.2.3 Stratification

In this section we introduce a notion of *stratification* relative to a given size function and some related terminology. Stratification provides important tools to study asymptotic behavior of subsets of inputs.

Let $s : I \to \mathbb{N}$ be a size function on a set of inputs $I$, satisfying the conditions C1), C2) from Section 3.2.2. For a given $n \in \mathbb{N}$ the set of elements of size $n$ in $I$,

$$I_n = \{x \in I \mid s(x) = n\},$$

is called the *sphere* of radius $n$ (or $n$-*stratum*), whereas the set

$$B_n(I) = \bigcup_{k=1}^{n} I_k = \{x \in I \mid s(x) \leq n\}$$

is called the *ball* of radius $n$.

The partition

$$I = \bigcup_{k=1}^{\infty} I_k \tag{3.1}$$

is called *size stratification* of $I$, and the decomposition

$$I = \bigcup_{k=1}^{\infty} B_k(I)$$

is the *size volume decomposition*. The converse also holds, i.e., every partition (3.1) induces a size function on $I$ such that $x \in I$ has size $k$ if and only if $x \in I_k$. In this case the condition C1) holds if and only if the partition (3.1) is computable, whereas C2) holds if and only if the sets $I_k$ are finite (or measurable, if the set $I$ comes equipped with a measure).

Size stratifications and volume decompositions allow one to study asymptotic behavior of various subsets of $I$, as well as some functions defined on some subsets of $I$. For example, if the size function $s : I \to \mathbb{N}$ is such that for any $n \in \mathbb{N}$ the set $I_n$ is finite (see condition C1), then for a given subset $R \subseteq I$ the function $n \to |R \cap I_n|$ is called the *growth* function of $R$ (relative to the given size stratification), while the function

$$n \to \rho_n(R) = \frac{|R \cap I_n|}{|I_n|}$$

is called the *frequency* function of $R$. The asymptotic behavior of $R$ can be seen via its *spherical asymptotic density*, which is defined as the following limit (if it exists):

$$\rho(R) = \lim_{n \to \infty} \rho_n(R).$$

Similarly, one can define the *volume asymptotic density* of $R$ as the limit (if it exists) of the *volume frequencies*:

$$\rho^*(R) = \lim_{n \to \infty} \rho_n^*(R),$$

where

$$\rho_n^*(R) = \frac{|R \cap B_n(I)|}{|B_n(I)|}.$$

One can also define the density functions above using $\limsup$ rather than $\lim$. We will have more to say about this in Section 8.1.2.

These spherical and volume stratifications play an important role in the study of complexity of algorithms and algorithmic problems.

### 3.2.4   Reductions and complete problems

In this section we discuss what it means when we say that one problem is as hard as another. A notion of *reduction* is used to define this concept. Intuitively, if a problem $D_1$ is reducible to a problem $D_2$, then via some "reduction procedure" a decision algorithm for $D_2$ gives a decision algorithm to $D_1$. This allows one to estimate the hardness of the problem $D_1$ through the hardness of $D_2$ and through the hardness of the reduction procedure.

One of the most general type of reductions is the so-called *Turing reduction* (see Section 3.2.6). It comes from the classical recursion theory and fits well for studying undecidable problems. Below we give a general description of one of the most common reductions, *many-to-one reduction* (see Section 3.2.5).

Let $F$ be a set of functions from $\mathbb{N}$ to $\mathbb{N}$, and $D_1$ and $D_2$ some decision problems (say, subsets of $\mathbb{N}$). The problem $D_1$ is called *reducible* to $D_2$ under $F$ if there exists a function $f \in F$ satisfying

$$x \in D_1 \iff f(x) \in D_2.$$

In this case we write $D_1 \leq_F D_2$. It is natural to assume that the set $F$ contains all identity functions and hence any problem $D$ is reducible to itself. Moreover, usually the set $F$ is closed under compositions which implies that if $D_1 \leq_F D_2$ and $D_2 \leq_F D_3$, then $D_1 \leq_F D_3$. Thus, $\leq_F$ is reflexive and transitive.

Now let $S$ be a set of decision problems and $\leq_F$ a reducibility relation. We say that $S$ is *closed under reductions* $\leq_F$ if for any problem $D_1$ and a problem $D_2 \in S$, one has

$$D_1 \leq_F D_2 \iff D_1 \in S.$$

A problem $C$ is called *hard* for $S$ if for any $D \in S$ we have $D \leq_F C$. A problem $C$ is called *complete* for $S$ if it is hard for $S$ and $C \in S$.

### 3.2.5 Many-one reductions

Let $D_1$ and $D_2$ be decision problems. A recursive function $f : D_1 \rightarrow D_2$ is called a *many-to-one reduction* of $D_1$ to $D_2$ if $x \in D_1$ if and only if $f(x) \in D_2$. In this case we say that $D_1$ is many-one reducible or *m-reducible* to $D_2$ and write $D_1 \leq_m D_2$. If $f$ is an injective many-one reduction, then we say that $D_1$ is one-to-one reducible to $D_2$ and write $D_1 \leq_1 D_2$.

Many-one reductions are a special case and a weaker form of Turing reductions. With many-one reductions only one invocation of the oracle is allowed, and only at the end.

Recall that a class $S$ of problems is closed under many-one reducibility if there are no many-one reductions from a problem in $S$ to a problem outside $S$. If a class $S$ is closed under many-one reducibility, then many-one reductions can be used to show that a problem is in $S$ by reducing a problem in $S$ to it.

The class of all decision problems has no complete problem with respect to many-one reductions.

### 3.2.6 Turing reductions

In this section we discuss the most general type of reductions called *Turing reductions*. Turing reductions are very important in computability theory and in providing theoretical grounds for more specific types of reductions, but like many-one reductions discussed in the previous section, they are of limited practical significance.

We say that a problem $D_1$ is *Turing reducible* to a problem $D_2$ and write $D_1 \leq_T D_2$ if $D_1$ is computable by a Turing machine with an oracle for $D_2$. The complexity class of decision problems solvable by an algorithm in class A with an oracle for a problem in class B is denoted by $A^B$. For example, the class of problems solvable in polynomial time by a deterministic Turing machine with an oracle for a problem in **NP** is $\mathbf{P^{NP}}$. (This is also the class of problems reducible by a polynomial-time Turing reduction to a problem in **NP**.)

It is possible to postulate the existence of an oracle which computes a non-computable function, such as an answer to the halting problem. A machine with an oracle of this sort is called a *hypercomputer*. Interestingly, the halting paradox still applies to such machines; that is, although they can determine whether particular Turing machines will halt on particular inputs, they cannot determine whether machines with equivalent halting oracles will themselves halt. This fact creates a hierarchy of machines, called the *arithmetical hierarchy*, each with a more powerful halting oracle and an even harder halting problem.

## 3.3   The worst case complexity

### 3.3.1   Complexity classes

We follow the book *Computational Complexity* by C. Papadimtriou [114] for our conventions on computational complexity. A *complexity class* is determined by specifying a *model of computation* (which for us is almost always a Turing machine), a *mode of computation* (e.g., deterministic or non-deterministic), *resources* to be controlled (e.g., time or space) and *bounds* for each controlled resource, which is a function $f$ from nonnegative integers to nonnegative integers satisfying certain properties as discussed below. The complexity class is now defined as the set of all languages decided by an appropriate Turing machine $M$ operating in the appropriate mode and such that, for any input $x$, $M$ uses at most $f(|x|)$ units of the specified resource. There are some restrictions on the bound function $f$. Speaking informally, we want to exclude those functions that require more resources to be computed than to read an input $n$ and an output $f(n)$.

**Definition 3.3.1.** Let $f$ be a function from nonnegative integers to nonnegative integers. We say that $f$ is a *proper complexity function* if the following conditions are satisfied:

1. $f$ is non-decreasing, i.e., for any $n \in \mathbb{N}$ $f(n + 1) \geq f(n)$.

2. There exists a multitape Turing machine $M_f$ such that for any $n$ and for any input $x$ of length $n$, $M$ computes a string $0^{f(|x|)}$ in time $T_M(x) = O(n + f(n))$ and space $S_M(x) = O(f(n))$.

The class of proper complexity bounds is very extensive and excludes mainly pathological cases of functions. In particular, it contains all polynomial functions, functions $\sqrt{n}$, $n!$, and $\log n$. Moreover, with any functions $f$ and $g$ it contains the functions $f + g$, $f \cdot g$, and $2^f$.

For a proper complexity function $f$, we define **TIME**$(f)$ to be the class of all decision problems decidable by some deterministic Turing machine within time $f(n)$. Then, let **NTIME**$(f)$ be the class of all decision problems decidable by some non-deterministic Turing machine within time $f(n)$. Similarly, we can define the complexity classes **SPACE**$(f)$ (deterministic space) and **NSPACE**$(f)$ (non-deterministic space).

Often classes are defined not for particular complexity functions but for parameterized families of functions. For instance, two most important classes of decision problems, **P** and **NP**, are defined as follows:

$$\mathbf{P} = \bigcup_{k=1}^{\infty} \mathbf{TIME}(n^k),$$

$$\mathbf{NP} = \bigcup_{k=1}^{\infty} \mathbf{NTIME}(n^k).$$

Other important classes are **PSPACE** $= \bigcup_{k=1}^{\infty} \textbf{SPACE}(n^k)$ (polynomial deterministic space) and **NPSPACE** $= \bigcup_{k=1}^{\infty} \textbf{NSPACE}(n^k)$ (polynomial nondeterministic space), **EXP** $= \bigcup_{k=1}^{\infty} \textbf{TIME}((2^n)^k)$ (exponential deterministic time).

As we have mentioned above, the principal computational models for us are Turing machines (in all their variations). However, we sometimes may describe them not by Turing machine programs, but rather less formally, as they appear in mathematics. Moreover, we may use other sequential models of computation, keeping in mind that we can always convert them into Turing machines, if needed. We refer to all these different descriptions and different models of computation as *algorithms*.

The complexity classes defined above are the so-called *worst case complexity* classes.

## 3.3.2 Class NP

It is a famous open problem whether $\textbf{P} = \textbf{NP}$ or not. There are several equivalent definitions of **NP**. According to our definition, a language $L \subseteq \{0,1\}^*$ belongs to **NP** if there exists a polynomial $p : \mathbb{N} \to \mathbb{N}$ and a polynomial-time TM $M$ such that for every $x \in \{0,1\}^*$,

$$x \in L \text{ if and only if } \exists u \in \{0,1\}^{p(|x|)} \text{ s.t. } M(x,u) = 1.$$

If $x \in L$ and $u \in \{0,1\}^{p(|x|)}$ satisfies $M(x,u) = 1$ then we say that $u$ is a *certificate* for $x$ (with respect to the language $L$ and machine $M$).

**Definition 3.3.2.** A Boolean formula $\varphi$ over variables $u_1, \ldots, u_n$ is in *conjunctive normal form* (CNF) if it is of the form

$$\wedge(\vee v_{i_j}),$$

where each $v_{i_j}$ is a literal of $\varphi$, in other words either a variable $u_k$ or its negation $\bar{u}_k$. The terms $\vee v_{i_j}$ are called *clauses*. If all clauses contain at most $k$ literals, the formula is called $k$-CNF.

Several problems in **NP**:

1. **Satisfiability problem, or SAT:** Given a boolean formula $\varphi$ in conjunctive normal form over variables $u_1, \ldots, u_n$, determine whether or not there is a satisfying assignment for $\varphi$.

2. **Three satisfiability problem, or 3SAT:** Given a boolean formula $\varphi$ in conjunctive normal form over variables $u_1, \ldots, u_n$, where each conjunct contains up to three variables, determine whether or not there is a satisfying assignment for $\varphi$.

3. **Subset sum problem:** Given a list of $n$ numbers $a_1, \ldots, a_n$ and a number $A$, decide whether or not there is a subset of these numbers that sums up to $A$. The certificate is the list of members in such a subset.

4. **Graph isomorphism:** Given two $n \times n$ adjacency matrices $M_1, M_2$, decide whether or not $M_1$ and $M_2$ define the same graph up to renaming of vertices. The certificate is a permutation $\pi \in S_n$ such that $M_1$ is equal to $M_2$ after reordering $M_1$'s entries according to $\pi$.

5. **Linear programming:** Given a list of $m$ linear inequalities with rational coefficients over $n$ variables $u_1, \ldots, u_n$ (a linear inequality has the form $a_1 u_1 + a_2 u_2 + \ldots a_n u_n \leq b$ for some coefficients $a_1, \ldots, a_n, b$), decide whether or not there is an assignment of rational numbers to the variables $u_1, \ldots, u_n$ that satisfies all the inequalities. The certificate is the assignment.

6. **Integer programming:** Given a list of $m$ linear inequalities with rational coefficients over $n$ variables $u_1, \ldots, u_n$, decide whether or not there is an assignment of integers to $u_1, \ldots, u_n$ that satisfies all the inequalities. The certificate is the assignment.

7. **Travelling salesperson:** Given a set of $n$ nodes, $\binom{n}{2}$ numbers $d_{i,j}$ denoting the distances between all pairs of nodes, and a number $k$, decide whether or not there is a closed circuit (i.e., a "salesperson tour") that visits every node exactly once and has total length at most $k$. The certificate is the sequence of nodes in the tour.

8. **Bounded halting problem:** Let $M$ be a non-deterministic Turing machine with binary input alphabet. The bounded halting problem for $M$ (denoted by $H(M)$) is the following algorithmic problem. For a positive integer $n$ and a binary string $x$, decide whether or not there is a halting computation of $M$ on $x$ within at most $n$ steps. The certificate is the halting execution flow (sequence of states of $M$).

9. **Post correspondence problem:** Given a nonempty list

$$L = ((u_1, v_1), \ldots, (u_s, v_s))$$

of pairs of binary strings and a positive integer $n$ determine whether or not there is a tuple $i = (i_1, \ldots, i_k)$ in $\{1, \ldots, s\}^k$, where $k \leq n$, such that

$$u_{i_1} u_{i_2} \ldots u_{i_k} = v_{i_1} v_{i_2} \ldots v_{i_k}.$$

The certificate is the tuple $i$.

### 3.3.3    Polynomial-time many-one reductions and class NP

Many-one reductions are often subjected to resource restrictions, for example that the reduction function is computable in polynomial time or logarithmic space. If not chosen carefully, the whole hierarchy of problems in the class can collapse to just one problem. Given decision problems $D_1$ and $D_2$ and an algorithm $A$ which solves instances of $D_2$, we can use a many-one reduction from $D_1$ to $D_2$ to solve instances of $D_2$ in:

- the time needed for $A$ plus the time needed for the reduction;
- the maximum of the space needed for $A$ and the space needed for the reduction.

Recall that **P** is the class of decision problems that can be solved in polynomial time by a deterministic Turing machine and **NP** is the class of decision problems that can be solved in polynomial time by a non-deterministic Turing machine. Clearly **P** $\subseteq$ **NP**. It is a famous open question (see e.g., [70]) whether **NP** $\subseteq$ **P** or not.

**Definition 3.3.3.** Let $D_1$ and $D_2$ be decision problems. We say that $f : D_1 \to D_2$ *Ptime reduces* $D_1$ to $D_2$ and write $D_1 \leq_P D_2$ if

- $f$ is polynomial time computable;
- $x \in D_1$ if and only if $f(x) \in D_2$.

We say that a Ptime reduction $f$ is *size-preserving* if

- $|x_1| < |x_2|$ if and only if $|f(x_1)| < |f(x_2)|$.

Recall that 3SAT is the following decision problem. Given a Boolean expression in conjunctive normal form, where each clause contains at most three variables, find out whether or not it is satisfiable. It is easy to check that 3SAT belongs to **NP**. The following theorem is a classical result.

**Theorem 3.3.4.** *The following is true:*

1) **NP** *is closed under Ptime reductions.*

2) *If $f$ is a Ptime reduction from $D_1$ to $D_2$ and $M$ is a Turing machine solving $D_2$ in polynomial time, then the composition of $f$ and $M$ solves $D_1$ in polynomial time.*

3) 3SAT *is complete in* **NP** *with respect to Ptime reductions.*

It follows from Theorem 3.3.4 that to prove (or disprove) that **NP** $\subseteq$ **P** one does not have to check all problems from **NP**; it is sufficient to show that $C \in$ **P** (or $C \notin$ **P**) for any **NP**-complete problem $C$.

### 3.3.4 NP-complete problems

In Section 3.2.4 we defined a notion of a hard and complete problem for a class of decision problems. In this section we use it to define **NP**-complete problems. Recall that a decision problem $L \subseteq \{0, 1\}^*$ belongs to the class **NP** if membership in $L$ has a polynomial-time verifiable certificate (witness). Now a problem $D$ is called **NP**-hard if for every problem $D' \in$ **NP** there exists a polynomial-time reduction of $D'$ to $D$. A problem is said to be **NP**-complete if it belongs to **NP** and is **NP**-hard.

**Theorem 3.3.5.** *There exists a nondeterministic* TM $U$ *such that* $H(U)$ *is* **NP**-*complete.*

*Proof.* (A sketch.) As we mentioned in Section 3.3.2, for any polynomial-time TM $U$, one has $H(U) \in \mathbf{NP}$. On the other hand, $D \in \mathbf{NP}$ if and only if there exists a polynomial-time NTM $M$ deciding $D$. Let $p(n)$ be the time-complexity of $M$. Fix an arbitrary problem $D$ from **NP**. Let $U$ be a universal NTM (which simulates execution flow of any TM $M$ on any input $x$) such that:

(1) $U$ takes inputs of the form $1^m 0 x$, where $m$ is a Gödel number of a TM $M$ to be simulated and $x$ is an input for $M$.

(2) $U$ is an efficient universal Turing machine, i.e., there exists a polynomial $q(n)$ for $U$ such that $M$ stops on $x$ in $k$ steps if and only if $U$ stops on $1^{m_M} 0 x$ in $q(k)$ steps.

Clearly, a function $f$ which maps an input $x$ for $M$ to the input $(q(p(|x|)), 1^m 0 x)$ for $H(U)$ is a polynomial-time reduction. Thus, $H(U)$ is **NP**-hard. $\square$

Let $M$ be a TM. We say that $M$ is oblivious if its head movement depends only on the length of input as a function of steps. It is not difficult to see that for any Turing machine $M$ with time complexity $T(n)$ there exists an oblivious TM $M'$ computing the same function as $M$ does with time complexity $T^2(n)$.

To prove the next theorem we need to introduce a notion of a snapshot for execution of a Turing machine. Assume that $M = \langle Q, \Sigma, s, f, \delta \rangle$ is a TM with two tapes. The first tape is read-only and is referred to as an *input tape*, and the second tape is in the read-write mode and is referred to as the *working tape*. A *snapshot* of $M$'s execution on input $y$ at step $i$ is a triple $(a, b, q) \in \Sigma \times \Sigma \times Q$, where $a$ and $b$ are the symbols observed by the the the $M$'s head and $q$ is the current state.

**Theorem 3.3.6.** SAT *is* **NP**-*complete*

*Proof.* SAT clearly belongs to **NP**, hence it remains to show that it is **NP**-hard. Let $D$ be an arbitrary **NP** problem. Our goal is to construct a polynomial time reduction $f$ which maps instances $x$ of $D$ to CNF formulas $\varphi_x$ such that $x \in D$ if and only if $\varphi_x$ is satisfiable. By definition, $D \in \mathbf{NP}$ if and only if there exists a polynomial $p(n)$ and a polynomial-time 2-tape TM $M = \langle Q, \Sigma, s, f, \delta \rangle$ such that for every $x \in \{0, 1\}^*$,

$$x \in D \text{ if and only if } \exists u \in \{0, 1\}^{p(|x|)} \text{ s.t. } M(x, u) = 1.$$

As mentioned above, we may assume that $M$ is oblivious (i.e., its head movement does not depend on the contents of the input tape, but only on the length of the input). The advantage of this property is that there are polynomial-time computable functions:

1. *inputpos*$(i)$ – the location of the input tape head at step $i$;

2. $prev(i)$ – the last step before $i$ that $M$ visited the same location on the work tape.

Each snapshot of $M$'s execution can be encoded by a binary string of length $c$, where $c$ is a constant depending on parameters of $M$, but independent of the input length.

For every $m \in \mathbb{N}$ and $y \in \{0,1\}^m$, the snapshot of $M$'s execution on the input $y$ at the $i$th step depends on its state at the $(i-1)$th step and on the contents of the current cells of its input and work tapes. Thus if we denote the encoding of the $i$th snapshot as a string of length $c$ by $z_i$, then $z_i$ is a function of $z_{i-1}$, $y_{inputpos(i)}$, and $z_{prev(i)}$, i.e.,

$$z_i = F(z_{i-1}, y_{inputpos(i)}, z_{prev(i)}),$$

where $F$ is some function that maps $\{0,1\}^{2c+1}$ to $\{0,1\}^c$.

Let $n \in \mathbb{N}$ and $x \in \{0,1\}^n$. We need to construct a CNF $\varphi_x$ such that $x \in D$ if and only if $\varphi_x \in SAT$. Recall that $x \in D$ if and only if there exists $u \in \{0,1\}^{p(n)}$ such that $M(y) = 1$, where $y = x \circ u$. Since the sequence of snapshots of $M$'s execution completely determines its outcome, this happens if and only if there exists a string $y \in \{0,1\}^{n+p(n)}$ and a sequence of strings $z_1, \ldots, z_T \in \{0,1\}^c$, where $T = T(n)$ is the number of steps $M$ takes on input of length $n + p(n)$ satisfying the following four conditions:

(1) The first $n$ bits of $y$ are the same as in $x$.

(2) The string $z_1$ encodes the initial snapshot of $M$.

(3) For every $i = 2, \ldots, T$, $z_i = F(z_{i-1}, y_{inputpos(i)}, z_{prev(i)})$.

(4) The last string $z_T$ encodes a snapshot in which the machine halts and outputs 1.

The formula $\varphi_x$ takes variables $y \in \{0,1\}^{n+p(n)}$ and $z \in \{0,1\}^{cT}$ and verifies that $y, z$ satisfy all of these four conditions. Clearly, $x \in D$ if and only if $\varphi_x \in SAT$, so it remains to show that we can express $\varphi_x$ as a polynomial-size CNF formula.

Condition (1) can be expressed as a CNF formula of size $2n$. Conditions (2) and (4) each depend on $c$ variables and hence can be expressed by a CNF formula of size $c2^c$. Condition (3) is an AND of $T$ conditions each depending on at most $3c + 1$ variables, hence it can be expressed as a CNF formula of size at most $T(3c + 1)2^{3c+1}$. Hence, AND of all these conditions can be expressed as a CNF formula of size $d(n + T)$, where $d$ is a constant depending only on $M$. Moreover, this CNF formula can be computed in polynomial time in the running time of $M$. $\square$

### 3.3.5 Deficiency of the worst case complexity

The worst case complexity of algorithms tells something about resources spent by a given algorithm $\mathcal{A}$ on the "worst possible" inputs for $\mathcal{A}$. Unfortunately, when the worst case inputs are sparse, this type of complexity tells nothing about the actual behavior of the algorithm on "most" or "most typical" (*generic*) inputs. However, what really counts in many practical applications is the behavior of algorithms on typical, most frequently occurring, inputs. For example, Dantzig's simplex algorithm for linear programming problems is used thousands of times daily and in practice almost always works quickly. It was shown by V. Klee and G. Minty [82] that there are some inputs where the simplex algorithm requires exponential time to finish computation, so it has exponential worst case time complexity. In [80] Khachiyan came up with a new ingenious algorithm for linear programming problems, that has polynomial time complexity. Nevertheless, at present Dantzig's algorithm continues to be most widely used in applications. The reason is that a "generic", or "random", linear programming problem is not "special", and the simplex algorithm works quickly. Mathematical justification of this phenomenon have been found independently by Vershik and Sporyshev [140] and Smale [130]. They showed that on a set of problems of measure 1, Dantzig's simplex algorithm has linear time complexity.

Modern cryptography is another area where computational complexity plays a crucial role because modern security assumptions in cryptography require analysis of behavior of algorithms on random inputs. The basic computational problem, which is in the core of a given cryptoscheme, must be hard on *most* inputs, to make it difficult for an attacker to crack the system. In this situation the worst case analysis of algorithms is irrelevant.

Observations of this type have led to the development of new types of complexity measure, where computational problems come equipped with a fixed distribution on the set of inputs (so called *distributional*, or *randomized*, computational problems). This setting allows one to describe the behavior of algorithms either on average or on "most" inputs. We discuss these complexity measures in more detail in Part III of our book.

The object of this part of the book is to explore how to use the complexity of non-commutative groups in *public-key cryptography*.

The idea of using the complexity of infinite, nonabelian groups in cryptography goes back to Magyarik and Wagner [97] who in 1985 devised a public-key protocol based on the unsolvability of the word problem for finitely presented groups (or so they thought). Their protocol now looks somewhat naive, but it was pioneering. More recently, there has been an increased interest in applications of nonabelian group theory to cryptography (see for example [1, 84, 129]). Most suggested protocols are based on *search problems* which are variants of more traditional *decision problems* of combinatorial group theory. Protocols based on search problems fit in with the general paradigm of a public-key protocol based on a one-way function (see above). We therefore dub the relevant area of cryptography *canonical cryptography* and explore it in Chapter 4 of our book.

On the other hand, employing decision problems in public-key cryptography allows one to depart from the canonical paradigm and construct cryptographic protocols with new properties, impossible in the canonical model. In particular, such protocols can be secure against some "brute force" attacks by computationally unbounded adversary. There is a price to pay for that, but the price is reasonable: a legitimate receiver decrypts correctly with probability that can be made arbitrarily close to 1, but not equal to 1. We discuss this and other new ideas in Chapter 11.

In a separate Chapter 5 we describe groups that can be used as platforms for cryptographic protocols from Chapters 4 and 11 of this book. These include braid groups, groups of matrices, small cancellation groups, and others.

# Chapter 4

# Canonical Non-commutative Cryptography

In this chapter, we discuss various cryptographic primitives that use non-commutative (semi)groups as platforms, but at the same time do not depart from the canonical paradigm of a public-key protocol based on a one-way function. We include here the ground-breaking Anshel-Anshel-Goldfeld protocol [1] as well as protocols that are closer in spirit to classical protocols based on commutative (semi)groups.

## 4.1  Protocols based on the conjugacy search problem

Let $G$ be a group with solvable word problem. For $w, a \in G$, the notation $w^a$ stands for $a^{-1}wa$. Recall that the *conjugacy problem* (or *conjugacy decision problem*) for $G$ is: given two elements $u, v \in G$, find out whether there is $x \in G$ such that $u^x = v$. On the other hand, the *conjugacy search problem* (sometimes also called the *witness conjugacy problem*) is: given two elements $a, b \in G$ and the information that $u^x = v$ for some $x \in G$, find at least one particular element $x$ like that.

The conjugacy decision problem is of great interest in group theory. In contrast, the conjugacy search problem is of interest in complexity theory, but of no interest in group theory. Indeed, if you know that $u$ is conjugate to $v$, you can just go over words of the form $u^x$ and compare them to $v$ one at a time, until you get a match. (We implicitly use here an obvious fact that a group with solvable conjugacy problem also has solvable word problem.) This straightforward algorithm is at least exponential-time in the length of $v$, and therefore is considered infeasible for practical purposes.

Thus, if no other algorithm is known for the conjugacy search problem in a group $G$, it is not unreasonable to claim that $x \to u^x$ is a one-way function and build a (public-key) cryptographic protocol on that.

We start with a simple protocol, due to Ko, Lee et. al. [84].

1. An element $w \in G$ is published.

2. Alice picks a private $a \in G$ and sends $w^a$ to Bob.

3. Bob picks a private $b \in G$ and sends $w^b$ to Alice.

4. Alice computes $(w^b)^a = w^{ba}$, and Bob computes $(w^a)^b = w^{ab}$.

If $a$ and $b$ are chosen from a pool of commuting elements of the group $G$, then $ab = ba$, and therefore, Alice and Bob get a common private key $w^{ab} = w^{ba}$. Typically, there are two public subgroups $A$ and $B$ of the group $G$, given by their (finite) generating sets, such that $ab = ba$ for any $a \in A$, $b \in B$.

In the paper [84], the platform group $G$ was the braid group $B_n$ which has some natural commuting subgroups. Selecting a suitable platform group for the above protocol is a very nontrivial matter; some requirements on such a group were put forward in [119]:

(P0) The group has to be well known. More precisely, the conjugacy search problem in the group either has to be well studied or can be reduced to a well-known problem (perhaps, in some other area of mathematics).

This property, although not mathematical, appears to be mandatory if we want our cryptographic product to be used in real life. We note in passing that this property already narrows down the list of candidates quite a bit.

(P1) The word problem in $G$ should have a fast (linear- or quadratic-time) solution by a deterministic algorithm. Better yet, there should be an efficiently computable "normal form" for elements of $G$.

This is required for an efficient common key extraction by legitimate parties in a key establishment protocol, or for the verification step in an authentication protocol, etc.

(P2) The conjugacy search problem should *not* have a subexponential-time solution by a deterministic algorithm.

We point out here that *proving* a group to have (P2) should be extremely difficult, if not impossible. This is, literally, a million-dollar problem (see [70]). The property (P2) should be therefore considered in conjunction with (P0), i.e., the only realistic evidence of a group $G$ having the property (P2) can be the fact that sufficiently many people have been studying the conjugacy search problem in $G$ over sufficiently long time.

The next property is somewhat informal, but it is of great importance for practical implementations:

(P3) There should be a way to disguise elements of $G$ so that it would be impossible to recover $x$ from $x^{-1}wx$ just by inspection.

One way to achieve this is to have a *normal form* for elements of $G$, which usually means that there is an algorithm that transforms any input

$u_{in}$, which is a word in the generators of $G$, to an output $u_{out}$, which is another word in the generators of $G$, such that $u_{in} = u_{out}$ in the group $G$, but this is hard to detect by inspection.

In the absence of a normal form, say if $G$ is just given by means of generators and relations without any additional information about properties of $G$, then at least some of these relations should be very short.

(P4) $G$ should be a group of super-polynomial (i.e., exponential or "intermediate") growth. This means that the number of elements of length $n$ in $G$ should grow faster than any polynomial in $n$; this is needed to prevent attacks by complete exhaustion of the key space. Here "length $n$" is typically just the length of a word representing a group element, but in a more general situation, this could be the length of some other description, i.e., "information complexity".

There are groups that have (P1), (P4), most likely have (P2), and to a reasonable extent have (P3). These are groups with solvable word problem, but unsolvable conjugacy problem (see e.g., [108]). However, the conjugacy search problem is of no independent interest in group theory, as we have mentioned in the beginning of this section. Group theorists therefore did not bother to study the conjugacy search problem in these groups once it had been proved that the conjugacy problem is algorithmically unsolvable. It would probably make sense to reconsider these groups now.

## 4.2 Protocols based on the decomposition problem

One of the natural ramifications of the conjugacy search problem is the following *decomposition search problem*:

> Given two elements $w$ and $w'$ of a group $G$, find two elements $x$ and $y$ that would belong to a given subset (usually a subgroup) $A \subseteq G$ and satisfy $x \cdot w \cdot y = w'$, provided at least one such pair of elements exists.

We note that if in the above problem $A$ is a subgroup, then this problem is also known as the *double coset problem* .

We also note that *some* $x$ and $y$ satisfying the equality $x \cdot w \cdot y = w'$ always exist (e.g., $x = 1$, $y = w^{-1}w'$), so the point is to have them satisfy the condition $x, y \in A$. We therefore will not usually refer to this problem as a *subgroup-restricted* decomposition search problem because it is always going to be subgroup-restricted; otherwise it does not make much sense.

We are going to show in Section 4.7 that solving the conjugacy search problem is unnecessary for an adversary to get the common secret key in the Ko-Lee (or any similar) protocol (see our Section 4.1); it is sufficient to solve a seemingly easier decomposition search problem. This was mentioned, in passing, in the paper [84], but the significance of this observation was downplayed there.

We note that the condition $x, y \in A$ may not be easy to verify for some subsets $A$; the corresponding problem is known as the membership (decision) problem. The authors of [84] do not address this problem; instead they mention, in justice, that if one uses a "brute force" attack by simply going over elements of $A$ one at a time, the above condition will be satisfied automatically. This however may not be the case with other, more practical, attacks.

We also note that the conjugacy search problem is a special case of the decomposition problem where $w'$ is conjugate to $w$ and $x = y^{-1}$. The claim that the decomposition problem should be easier than the conjugacy search problem is intuitively clear since it is generally easier to solve an equation with two unknowns than a special case of the same equation with just one unknown. We admit however that there might be exceptions to this general rule.

Now we give a formal description of a typical protocol based on the decomposition problem. There is a public group $G$, and two public subgroups $A, B \subseteq G$ commuting elementwise, i.e., $ab = ba$ for any $a \in A$, $b \in B$.

1. Alice randomly selects private elements $a_1, a_2 \in A$. Then she sends the element $a_1 w a_2$ to Bob.

2. Bob randomly selects private elements $b_1, b_2 \in B$. Then he sends the element $b_1 w b_2$ to Alice.

3. Alice computes $K_A = a_1 b_1 w b_2 a_2$, and Bob computes $K_B = b_2 a_1 w b_1 a_2$. Since $a_i b_i = b_i a_i$ in $G$, one has $K_A = K_B = K$ (as an element of $G$), which is now Alice's and Bob's common secret key.

We now discuss several modifications of the above protocol.

## 4.2.1   "Twisted" protocol

This idea is due to Shpilrain and Ushakov [122]; the following modification of the above protocol appears to be more secure (at least for some choices of the platform group) against so-called "length based" attacks (see e.g., [65], [67]), according to computer experiments. Again, there is a public group $G$, and two public subgroups $A, B \leq G$ commuting elementwise.

1. Alice randomly selects private elements $a_1 \in A$ and $b_1 \in B$. Then she sends the element $a_1 w b_1$ to Bob.

2. Bob randomly selects private elements $b_2 \in B$ and $a_2 \in A$. Then he sends the element $b_2 w a_2$ to Alice.

3. Alice computes $K_A = a_1 b_2 w a_2 b_1 = b_2 a_1 w b_1 a_2$, and Bob computes $K_B = b_2 a_1 w b_1 a_2$. Since $a_i b_i = b_i a_i$ in $G$, one has $K_A = K_B = K$ (as an element of $G$), which is now Alice's and Bob's common secret key.

The security of this protocol is apparently based on a more general version of the decomposition search problem than that given in the beginning of the previous subsection:

> *Given two elements $w$, $w'$ and two subgroups $A, B$ of a group $G$, find two elements $x \in A$ and $y \in B$ such that $x \cdot w \cdot y = w'$, provided at least one such pair of elements exists.*

## 4.2.2 Hiding one of the subgroups

This idea, too, is due to Shpilrain and Ushakov [124]. First we give a sketch of the idea.

Let $G$ be a group and $g \in G$. Denote by $C_G(g)$ the *centralizer* of $g$ in $G$, i.e., the set of elements $h \in G$ such that $hg = gh$. For $S = \{g_1, \ldots, g_k\} \subseteq G$, $C_G(g_1, \ldots, g_k)$ denotes the centralizer of $S$ in $G$, which is the intersection of the centralizers $C_G(g_i)$, $i = 1, \ldots, k$.

Now, given a public $w \in G$, Alice privately selects $a_1 \in G$ and publishes a subgroup $B \subseteq C_G(a_1)$ (we tacitly assume here that $B$ can be computed efficiently). Similarly, Bob privately selects $b_2 \in G$ and publishes a subgroup $A \subseteq C_G(b_2)$. Alice then selects $a_2 \in A$ and sends $w_1 = a_1 w a_2$ to Bob, while Bob selects $b_1 \in B$ and sends $w_2 = b_1 w b_2$ to Alice.

Thus, in the first transmission, say, the adversary faces the problem of finding $a_1, a_2$ such that $w_1 = a_1 w a_2$, where $a_2 \in A$, but there is no explicit indication of where to choose $a_1$ from. Therefore, before arranging something like a length attack in this case, the adversary would have to compute generators of the centralizer $C_G(B)$ first (because $a_1 \in C_G(B)$), which is usually a hard problem by itself since it basically amounts to finding the intersection of the centralizers of individual elements, and finding (the generators of) the intersection of subgroups is a notoriously difficult problem for most groups considered in combinatorial group theory.

Now we give a formal description of the protocol from [124]. As usual, there is a public group $G$, and let $w \in G$ be public, too.

1. Alice chooses an element $a_1 \in G$, chooses a subgroup of $C_G(a_1)$, and publishes its generators $A = \{\alpha_1, \ldots, \alpha_k\}$.

2. Bob chooses an element $b_2 \in G$, chooses a subgroup of $C_G(b_2)$, and publishes its generators $B = \{\beta_1, \ldots, \beta_m\}$.

3. Alice chooses a random element $a_2$ from $\langle \beta_1, \ldots, \beta_m \rangle$ and sends $P_A = a_1 w a_2$ to Bob.

4. Bob chooses a random element $b_1$ from $\langle \alpha_1, \ldots, \alpha_k \rangle$ and sends $P_B = b_1 w b_2$ to Alice.

5. Alice computes $K_A = a_1 P_B a_2$.

6. Bob computes $K_B = b_1 P_A b_2$.

Since $a_1 b_1 = b_1 a_1$ and $a_2 b_2 = b_2 a_2$, we have $K = K_A = K_B$, the shared secret key.

### 4.2.3  Using the triple decomposition problem

This protocol is due to Kurt [86]. Her idea is to base security of a key exchange primitive on the "triple decomposition problem", where a known element is to be factored into a product of three unknown factors. Recall that in the "usual" decomposition problem a known element is to be factored into a product of three factors, where two factors are unknown and one (in the middle) is known. The motivation for this modification is the following: if the platform group which is used with the corresponding protocol is linear, then the "usual" decomposition problem can be reduced to a system of linear equations, thus making a linear algebra attack possible. This is illustrated, in particular, in our Section 4.4. With the triple decomposition problem, a similar trick can only reduce the problem to a system of *quadratic* equations which is, of course, less malleable at least for standard attacks.

In Kurt's protocol the private key has three components. The idea is to hide each of these components by multiplying them by random elements of a (public) subgroup. The important part is that one of the components is multiplied by random elements both on the right and on the left. Now we are getting to the protocol description.

There is a public platform group (or a monoid) $G$ and two sets of subsets of $G$ containing five subsets of $G$ each, say $A = A_1, A_2, A_3, X_1, X_2$ and $B = B_1, B_2, B_3, Y_1, Y_2$, satisfying the following invertibility and commutativity conditions:

**(Invertibility conditions)** The elements of $X_1, X_2, Y_1, Y_2$ are invertible.

**(Commutativity conditions)** $[A_2, Y_1] = 1$, $[A_3, Y_2] = 1$, $[B_1, X_1] = 1$, and $[B_2, X_2] = 1$.

Alice and Bob agree on who will use which set of subsets; say Alice uses $A$ and Bob uses $B$. Then the exchange between Alice and Bob goes as follows:

1. Alice chooses $a_1 \in A_1$, $a_2 \in A_2$, $a_3 \in A_3$, $x_1 \in X_1$, $x_2 \in X_2$, and computes: $u = a_1 x_1$, $v = x_1^{-1} a_2 x_2$, and $w = x_2^{-1} a_3$. Her private key is $(a_1, a_2, a_3)$.

2. Bob chooses $b_1 \in B_1$, $b_2 \in B_2$, $b_3 \in B_3$, $y_1 \in Y_1$, $y_2 \in Y_2$, and computes: $p = b_1 y_1$, $q = y_1^{-1} b_2 y_2$, and $r = y_2^{-1} b_3$. His private key is $(b_1, b_2, b_3)$.

3. Alice sends $(u, v, w)$ to Bob.

4. Bob sends $(p, q, r)$ to Alice.

5. Alice computes $a_1 p a_2 q a_3 r = a_1 (b_1 y_1) a_2 (y_1^{-1} b_2 y_2) a_3 (y_2^{-1} b_3) = a_1 b_1 a_2 b_2 a_3 b_3$ $= K_A$.

6. Bob computes $u b_1 v b_2 w b_3 = (a_1 x_1) b_1 (x_1^{-1} a_2 x_2) b_2 (x_2^{-1}) a_3 b_3 = a_1 b_1 a_2 b_2 a_3 b_3$ $= K_B$.

Thus, $K_A = K_B = K$ is Alice's and Bob's common secret key. We refer to [86] for details on parameters of this scheme.

## 4.3 A protocol based on the factorization search problem

In this section, we describe a protocol based on the *factorization search problem*:

> Given an element $w$ of a group $G$ and two subgroups $A, B \leq G$, find any two elements $a \in A$ and $b \in B$ that would satisfy $a \cdot b = w$.

As before, there is a public group $G$, and two public subgroups $A, B \leq G$ commuting elementwise, i.e., $ab = ba$ for any $a \in A, b \in B$.

1. Alice randomly selects private elements $a_1 \in A$, $b_1 \in B$. Then she sends the element $a_1 b_1$ to Bob.

2. Bob randomly selects private elements $a_2 \in A$, $b_2 \in B$. Then he sends the element $a_2 b_2$ to Alice.

3. Alice computes:

$$K_A = b_1(a_2 b_2)a_1 = a_2 b_1 a_1 b_2 = a_2 a_1 b_1 b_2,$$

and Bob computes:

$$K_B = a_2(a_1 b_1)b_2 = a_2 a_1 b_1 b_2.$$

Thus, $K_A = K_B = K$ is now Alice's and Bob's common secret key.

We note that the adversary, Eve, who knows the elements $a_1 b_1$ and $a_2 b_2$, can compute $(a_1 b_1)(a_2 b_2) = a_1 b_1 a_2 b_2 = a_1 a_2 b_1 b_2$ and $(a_2 b_2)(a_1 b_1) = a_2 a_1 b_2 b_1$, but neither of these products is equal to $K$ if $a_1 a_2 \neq a_2 a_1$ and $b_1 b_2 \neq b_2 b_1$.

Finally, we point out a *decision* factorization problem:

> Given an element $w$ of a group $G$ and two subgroups $A, B \leq G$, find out whether or not there are two elements $a \in A$ and $b \in B$ such that $w = a \cdot b$.

This seems to be a new and nontrivial problem in group theory, and therefore it gives an example of a group-theoretic problem motivated by cryptography, i.e., we have here a remarkable feedback from cryptography to group theory.

## 4.4 Stickel's key exchange protocol

Stickel's protocol [133] is reminiscent of the classical Diffie-Hellman protocol (see our Section 1.2), although formally it is not a generalization of the latter. We show below, in Section 4.4.1, that Stickel's choice of platform (the group of invertible matrices over a finite field) makes the protocol vulnerable to linear algebra attacks. It appears however that even such a seemingly minor improvement as using non-invertible matrices instead of invertible ones would already make Stickel's protocol significantly less vulnerable, at least to linear algebra attacks, which is why we think this protocol has some potential. Our exposition in this section follows [121].

Let $G$ be a public nonabelian finite group, $a, b \in G$ public elements such that $ab \neq ba$. The key exchange protocol goes as follows. Let $N$ and $M$ be the orders of $a$ and $b$, respectively.

1. Alice picks two random natural numbers $n < N$, $m < M$ and sends $u = a^n b^m$ to Bob.

2. Bob picks two random natural numbers $r < N$, $s < M$ and sends $v = a^r b^s$ to Alice.

3. Alice computes $K_A = a^n v b^m = a^{n+r} b^{m+s}$.

4. Bob computes $K_B = a^r u b^s = a^{n+r} b^{m+s}$.

Thus, Alice and Bob end up with the same group element $K = K_A = K_B$ which can serve as the shared secret key.

When it comes to implementation details, exposition in [133] becomes somewhat foggy. In particular, it seems that the author actually prefers the following more general version of the above protocol.

Let $w \in G$ be public.

1. Alice picks two random natural numbers $n < N$, $m < M$, an element $c_1$ from the center of the group $G$, and sends $u = c_1 a^n w b^m$ to Bob.

2. Bob picks two random natural numbers $r < N$, $s < M$, an element $c_2$ from the center of the group $G$, and sends $v = c_2 a^r w b^s$ to Alice.

3. Alice computes $K_A = c_1 a^n v b^m = c_1 c_2 a^{n+r} w b^{m+s}$.

4. Bob computes $K_B = c_2 a^r u b^s = c_1 c_2 a^{n+r} w b^{m+s}$.

Thus, Alice and Bob end up with the same group element $K = K_A = K_B$.

We note that for this protocol to work, $G$ does not have to be a group; a semigroup would do just as well (in fact, even better, as we argue below).

In [133], it was suggested that the group of invertible $k \times k$ matrices over a finite field $F_{2^l}$ is used as the platform group $G$. We show in Section 4.4.1 that this choice of platform makes the protocol vulnerable to linear algebra attacks, but first we are going to discuss a general (i.e., not platform-specific) approach to attacking Stickel's protocol. We emphasize that this general approach works if $G$ is any semigroup, whereas the attack in Section 4.4.1 is platform-specific; in particular, it only works if $G$ is a group, but may not work for arbitrary semigroups.

Recall now that Alice transmits $u = c_1 a^n w b^m$ to Bob.

Our first observation is: to get a hold of the shared secret key $K$ in the end, it is sufficient for the adversary (Eve) to find any elements $x, y \in G$ such that $xa = ax$, $yb = by$, $u = xwy$. Indeed, having found such $x, y$, Eve can use Bob's transmission $v = c_2 a^r w b^s$ to compute:

$$xvy = xc_2 a^r w b^s y = c_2 a^r xwy b^s = c_2 a^r u b^s = K.$$

This implies, in particular, that multiplying by $c_i$ does not enhance security of the protocol. More importantly, this also implies that it is not necessary for Eve to recover any of the exponents $n, m, r, s$; instead, she can just solve a system of equations $xa = ax$, $yb = by$, $u = xwy$, where $a, b, u, w$ are known and $x, y$ unknown elements of the platform (semi)group $G$. This shows that, in fact, Stickel's protocol departs from the Diffie-Hellman protocol farther than it seems. Moreover, solving the above system of equations in $G$ is actually nothing else but solving the (subsemigroup-restricted) *decomposition search problem* (see our Section 2.3.3):

> Given a recursively presented (semi)group $G$, two recursively generated sub(semi)groups $A, B \leq G$, and two elements $u, w \in G$, find two elements $x \in A$ and $y \in B$ that would satisfy $x \cdot w \cdot y = u$, provided at least one such pair of elements exists.

In reference to Stickel's scheme, the sub(semi)groups $A$ and $B$ are the *centralizers* of the elements $a$ and $b$, respectively. The centralizer of an element $g \in G$ is the set of all elements $c \in G$ such that $gc = cg$. This set is a subsemigroup of $G$; if $G$ is a group, then this set is a subgroup.

So far, no particular (semi)group has been recognized as providing a secure platform for any of the protocols based on the decomposition search problem. It appears likely that semigroups of matrices over specific rings can generally make good platforms, see [120]. Stickel, too, used matrices in his paper [133], but he has made several poor choices, as we are about to see in the next Section 4.4.1. Also, Stickel's scheme is *at most* as secure as those schemes that are directly based on the alleged hardness of the decomposition search problem, because there are ways to attack Stickel's scheme without attacking the relevant decomposition search problem; for instance, Sramka [131] has offered an attack aimed at recovering one of the exponents $n, m, r, s$ in Stickel's protocol. Our attack that we describe in Section 4.4.1 is more efficient, but on the other hand, it is aimed at recovering the shared secret key only, whereas Sramka's attack is aimed at recovering a private key.

## 4.4.1 Linear algebra attack

Now we are going to focus on the particular platform group $G$ suggested by Stickel in [133]. In his paper, $G$ is the group of invertible $k \times k$ matrices over a finite field $F_{2^l}$, where $k = 31$. The parameter $l$ was not specified in [133], but from what is written there, one can reasonably guess that $2 \leq l \leq k$. The choice of matrices $a, b, w$ is not so important for our attack; what is important is that $a$ and $b$ are invertible. We note however that the choice of matrices $a$ and $b$ in [133] (more specifically, the fact that the entries of these matrices are either 0 or 1) provides an extra weakness to the scheme as we will see at the end of this section.

Recall that it is sufficient for Eve to find at least one solution of the system of equations $xa = ax$, $yb = by$, $u = xwy$, where $a, b, u, w$ are known and $x, y$ unknown $k \times k$ matrices over $F_{2^l}$. Each of the first two equations in this system translates into a system of $k^2$ linear equations for the (unknown) entries of the matrices $x$ and $y$. However, the equation $u = xwy$ does not translate into a system of linear equations for the entries because it has a product of two unknown matrices. We therefore have to use the following trick: multiply both sides of the equation $u = xwy$ by $x^{-1}$ on the left (here is where we use the fact that $x$ is invertible!) to get

$$x^{-1}u = wy.$$

Now, since $xa = ax$ if and only if $x^{-1}a = ax^{-1}$, we denote $x_1 = x^{-1}$ and replace the system of equations mentioned in the previous paragraph by the following one:

$$x_1 a = a x_1, \quad yb = by, \quad x_1 u = wy.$$

Now each equation in this system translates into a system of $k^2$ linear equations for the (unknown) entries of the matrices $x_1$ and $y$. Thus, we have a total of $3n^2$ linear equations with $2k^2$ unknowns. Note however that a solution of the displayed system will yield the shared key $K$ if and only if $x_1$ is invertible because $K = xvy$, where $x = x_1^{-1}$.

Since $u$ is a known invertible matrix, we can multiply both sides of the equation $x_1 u = wy$ by $u^{-1}$ on the right to get $x_1 = wyu^{-1}$, and then eliminate $x_1$ from the system:

$$wyu^{-1}a = awyu^{-1}, \quad yb = by.$$

Now we have just one unknown matrix $y$, so we have $2k^2$ linear equations for $k^2$ entries of $y$. Thus, we have a heavily overdetermined system of linear equations (recall that in Stickel's paper, $k = 31$, so $k^2 = 961$). We know that this system must have at least one nontrivial (i.e., nonzero) solution; therefore, if we reduce the matrix of this system to an echelon form, there should be at least one free variable. On the other hand, since the system is heavily overdetermined, we can expect the number of free variables to be not too big, so that it is feasible to go over possible values of free variables one at a time, until we find some values that yield an invertible matrix $y$. (Recall that entries of $y$ are either 0 or 1; this is an extra weakness of Stickel's scheme that we mentioned before.) Note that checking the invertibility of a given matrix is easy because it is equivalent to reducing the matrix to an echelon form. In fact, in all our experiments there was just one free variable, so the last step (checking the invertibility) was not needed because if there is a unique nonzero solution of the above system, then the corresponding matrix $y$ should be invertible.

Thus, the most obvious suggestion on improving Stickel's scheme is, as we mentioned before, to use non-invertible elements $a, b, w$; this implies, in particular, that the platform should be a semigroup with (a lot of) non-invertible elements. If one is to use matrices, then it makes sense to use the semigroup of *all* $k \times k$ matrices over a finite ring (not necessarily a field!). Such a semigroup typically has a lot of non-invertible elements, so it should be easy to choose $a, b, w$ non-invertible, in which case the linear algebra attack would not work. One more advantage of not restricting the pool to invertible matrices is that one can use not just powers $a^j$ of a given public matrix in Stickel's protocol, but arbitrary expressions of the form $\sum_{i=1}^{p} c_i \cdot a^i$, where $c_i$ are constants, i.e., elements of the ground ring.

Of course, there is no compelling reason why matrices should be employed in Stickel's scheme, but as we have explained above, with an abstract platform (semi)group $G$, Stickel's scheme is broken if the relevant decomposition search problem is solved, and so far, no particular abstract (semi)group has been recognized as resistant to known attacks on the decomposition search problem.

## 4.5 The Anshel-Anshel-Goldfeld protocol

In this section, we are going to describe a key establishment protocol that really stands out because, unlike other protocols in this chapter, it does not employ any commuting or commutative subgroups of a given platform group and can, in fact, use any nonabelian group with efficiently solvable word problem as the platform. This really makes a difference and gives a big advantage to the protocol of [1] over other protocols in this chapter.

A group $G$ and elements $a_1, \ldots, a_k, b_1, \ldots, b_m \in G$ are public.

(1) Alice picks a private $x \in G$ as a word in $a_1, \ldots, a_k$ (i.e., $x = x(a_1, \ldots, a_k)$) and sends $b_1^x, \ldots, b_m^x$ to Bob.

(2) Bob picks a private $y \in G$ as a word in $b_1, \ldots, b_m$ and sends $a_1^y, \ldots, a_k^y$ to Alice.

(3) Alice computes $x(a_1^y, \ldots, a_k^y) = x^y = y^{-1}xy$, and Bob computes $y(b_1^x, \ldots, b_m^x) = y^x = x^{-1}yx$. Alice and Bob then come up with a common private key $K = x^{-1}y^{-1}xy$ (called the *commutator* of $x$ and $y$) as follows: Alice multiplies $y^{-1}xy$ by $x^{-1}$ on the left, while Bob multiplies $x^{-1}yx$ by $y^{-1}$ on the left, and then takes the inverse of the whole thing: $(y^{-1}x^{-1}yx)^{-1} = x^{-1}y^{-1}xy$.

It may seem that solving the (simultaneous) conjugacy search problem for $b_1^x, \ldots, b_m^x; a_1^y, \ldots, a_k^y$ in the group $G$ would allow an adversary to get the secret key $K$. However, if we look at Step (3) of the protocol, we see that the adversary would have to know either $x$ or $y$ not simply as a word in the generators of the group $G$, but as a word in $a_1, \ldots, a_k$ (respectively, as a word in $b_1, \ldots, b_m$); otherwise, he would not be able to compose, say, $x^y$ out of $a_1^y, \ldots, a_k^y$. That means the adversary would also have to solve the *membership search problem*:

Given elements $x, a_1, \ldots, a_k$ of a group $G$, find an expression (if it exists) of $x$ as a word in $a_1, \ldots, a_k$.

We note that the membership *decision* problem is to determine whether or not a given $x \in G$ belongs to the subgroup of $G$ generated by given $a_1, \ldots, a_k$. This problem turns out to be quite hard in many groups. For instance, the membership decision problem in a braid group $B_n$ is algorithmically unsolvable if $n \geq 6$ because such a braid group contains subgroups isomorphic to $F_2 \times F_2$ (that would be, for example, the subgroup generated by $\sigma_1^2, \sigma_2^2, \sigma_4^2,$ and $\sigma_5^2$, see [23]), where $F_2$ is the free group of rank 2. In the group $F_2 \times F_2$, the membership decision problem is algorithmically unsolvable by an old result of Mihailova [107].

We also note that if the adversary finds, say, some $x' \in G$ such that $b_1^x = b_1^{x'}, \ldots, b_m^x = b_m^{x'}$, there is no guarantee that $x' = x$ in $G$. Indeed, if $x' = c_b x$, where $c_b b_i = b_i c_b$ for all $i$ (in which case we say that $c_b$ *centralizes* $b_i$), then $b_i^x = b_i^{x'}$ for all $i$, and therefore $b^x = b^{x'}$ for any element $b$ from the subgroup generated by $b_1, \ldots, b_m$; in particular, $y^x = y^{x'}$. Now the problem is that if $x'$ (and, similarly, $y'$) does not belong to the subgroup $A$ generated by $a_1, \ldots, a_k$ (respectively, to the subgroup $B$ generated by $b_1, \ldots, b_m$), then the adversary may not obtain the correct common secret key $K$. On the other hand, if $x'$ (and, similarly, $y'$) does belong to the subgroup $A$ (respectively, to the subgroup $B$), then the adversary will be able to get the correct $K$ even though his $x'$ and $y'$ may be different from $x$ and $y$, respectively. Indeed, if $x' = c_b x$, $y' = c_a y$, where $c_b$ centralizes $B$ and $c_a$ centralizes $A$ (elementwise), then

$$(x')^{-1}(y')^{-1}x'y' = (c_b x)^{-1}(c_a y)^{-1} c_b x c_a y = x^{-1} c_b^{-1} y^{-1} c_a^{-1} c_b x c_a y = x^{-1} y^{-1} xy$$
$$= K$$

because $c_b$ commutes with $y$ and with $c_a$ (note that $c_a$ belongs to the subgroup $B$, which follows from the assumption $y' = c_a y \in B$, and, similarly, $c_b$ belongs to $A$), and $c_a$ commutes with $x$.

We emphasize that the adversary ends up with the corrrect key $K$ (i.e., $K = (x')^{-1}(y')^{-1}x'y' = x^{-1}y^{-1}xy$) if and only if $c_b$ commutes with $c_a$. The only visible way to ensure this is to have $x' \in A$ and $y' \in B$. Without verifying at least one of these inclusions, there seems to be no way for the adversary to make sure that he got the correct key.

Therefore, it appears that if the adversary chooses to solve the conjugacy search problem in the group $G$ to recover $x$ and $y$, he will then have to face either the membership search problem or the membership decision problem; the latter may very well be algorithmically unsolvable in a given group. The bottom line is that the adversary should actually be solving a more difficult version of the conjugacy search problem:

Given a group $G$, a subgroup $A \leq G$, and two elements $g, h \in G$, find $x \in A$ such that $h = x^{-1}gx$, given that at least one such $x$ exists.

Finally, we note that what we have said in this section does not affect some heuristic attacks on the Anshel-Anshel-Goldfeld protocol suggested by several authors [42, 67] because these attacks, which use "neighborhood search" type (in a group-theoretic context also called "length based") heuristic algorithms, are targeted, by design, at finding a solution of a given equation (or a system of equations) as a word in given elements. The point that we make in this section is that even if a fast (polynomial-time) *deterministic* algorithm is found for solving the conjugacy search problem in braid groups, this will not be sufficient to break the Anshel-Anshel-Goldfeld protocol *by a deterministic attack*.

## 4.6 Authentication protocols based on the conjugacy problem

The main goal of an authentication protocol is to allow a legitimate user (Alice) to prove her identity over an insecure channel to a server (Bob) using her private key without leaking any information about the key. Alice is usually referred to as *prover* and Bob is referred to as *verifier*.

There are several general proposals available that use nonabelian groups; most of them are direct adaptations of key exchange schemes such as, e.g., the Ko-Lee key exchange protocol.

### 4.6.1 A Diffie-Hellman-like scheme

Let $G$ be a group and $A, B < G$ two commuting subgroups of $G$, i.e., $ab = ba$ for any $a \in A$ and $b \in B$.

Alice's private key is an element $s \in A$. Alice's public key is a pair $(w, t)$, where $w$ is an arbitrary element of $G$ and $t = s^{-1}ws$. The authentication protocol is the following sequence of transactions:

1. Bob chooses $r \in B$ and sends a challenge $w' = r^{-1}wr$ to Alice.

2. Alice sends the response $w'' = s^{-1}w's$ to Bob.

3. Bob checks if $w'' = r^{-1}tr$.

A correct answer of the prover at the second step leads to acceptance by the verifier because by design the elements $r$ and $s$ commute, and hence the equality

$$w'' = s^{-1}w's = s^{-1}r^{-1}wrs = r^{-1}s^{-1}wsr = r^{-1}tr$$

is satisfied.

The intruder (Eve) who wants to authenticate as Alice to Bob can do either of the following:

- **Compute Alice's private key** $s$ by solving the conjugacy search problem relative to the subgroup $A$ or by solving the decomposition problem for a pair $(w, t)$ relative to a subgroup $A$, i.e., computing elements $s_1, s_2 \in A$ such that $t = s_1ws_2$. It is trivial to check that for such a pair $(s_1, s_2)$ the equality $s_1w's_2 = w''$ is always satisfied and hence the pair $(s_1, s_2)$ works as Alice's private key $s$.

- **Compute Bob's temporary key** $r$ by solving the conjugacy search problem relative to the subgroup $B$ or by solving the decomposition problem for a pair $(w, w')$ relative to the subgroup $B$ and use it as described above for attack on Alice's private key.

The computational hardness of these problems depends on particular subgroups $A$ and $B$, so it does not have to be the same for $A$ and $B$.

**Remark 4.6.1.** This scheme can be easily modified to be based on the decomposition problem. Indeed, suppose that $A_1, A_2, B_1, B_2$ are subgroups of $G$ such that $[A_1, B_1] = 1$ and $[A_2, B_2] = 1$. Alice's private key is a pair $(a_1, a_2) \in A_1 \times A_2$, and her public key is a pair $(w, t) \in G \times G$, where $w$ is an arbitrary element of $G$ and $t = a_1 w a_2$. Now the transactions are modified as follows:

- Bob chooses elements $b_1 \in B_1$, $b_2 \in B_2$ and sends a challenge $w' = b_1 w b_2$ to Alice.

- Alice sends the response $w'' = a_1 w' a_2$ to Bob.

- Bob checks if $w'' = b_1 t b_2$.

This scheme is due to Lal and Chaturwedi [88]. It seems that the conjugacy search problem cannot be used to attack this modification of the protocol.

## 4.6.2   A Fiat-Shamir-like scheme

Another proposal of a group based authentication scheme is due to Sibert et. al. [128]. This scheme is reminiscent of the Fiat-Shamir scheme [37] which involves repeating several times a three-pass challenge-response step. One of the important differences from the Diffie-Hellman-like scheme described above is that there is no need to choose commuting subgroups $A$ and $B$.

As above, Alice's private key is an element $s \in G$ and her public key is a pair $(w, t)$, where $w$ is an arbitrary element of $G$ and $t = s^{-1} w s$. The authentication protocol goes as follows:

1. Alice chooses a random $r \in G$ and sends the element $x = r^{-1} t r$, called the *commitment*, to Bob.

2. Bob chooses a random bit $c$ and sends it to Alice.

   - If $c = 0$, then Alice sends $y = r$ to Bob and Bob checks if the equality $x = y^{-1} t y$ is satisfied.

   - If $c = 1$, then Alice sends $y = sr$ to Bob and Bob checks if the equality $x = y^{-1} w y$ is satisfied.

It is trivial to check that a correct answer of the prover at the second step leads to acceptance by the verifier. It is somewhat less obvious why such an arrangement (with a random bit) is needed; it may seem that Alice could just reveal $y = sr$: this does not reveal the secret $s$, and yet allows Bob to verify the equality $x = y^{-1} w y$.

The point is that in this case, the adversary, Eve, who wants to impersonate Alice can just take an arbitrary element $u$ and send $x = u^{-1}wu$ to Bob as a commitment. Then this $u$ would play the same role as $y = sr$ for verification purposes. Similarly, if Eve knew for sure that Alice would send $y = r$ for verification purposes, she would use an arbitrary element $u$ in place of $r$ at the commitment step. These considerations also show that the above authentication protocol should be run several times for better reliability because with a single run, Eve can successfully impersonate Alice with probability $\frac{1}{2}$. After $k$ runs, this probability goes down to $\frac{1}{2^k}$.

Similarly to the Diffie-Hellman-like scheme, the security of the Fiat-Shamir-like scheme depends on the computational hardness of the conjugacy search problem in the group $G$. It is interesting to note that the decomposition search problem cannot be used to attack this particular protocol because Bob accepts exactly one element, either $r$ or $sr$, depending on the value of $c$, at the last step of the protocol.

### 4.6.3 An authentication scheme based on the twisted conjugacy problem

The authentication scheme described in the previous section can be modified to be based on the decomposition problem in a similar way as that was done in Section 4.6.1. This scheme also allows a more interesting modification. Let $\varphi$ be an arbitrary endomorphism (i.e., homomorphism into itself) of the platform group $G$. We assume that $\varphi$ is publicly known, e.g., it can be part of Alice's public key. Alice's private key is an element $s \in G$, and her public key is a pair $(w, t)$, where $w$ is an arbitrary element of $G$ and $t = s^{-1}w\varphi(s)$. The authentication protocol goes as follows:

- Alice selects an element $r \in G$ and sends the element $x = r^{-1}t\varphi(r)$, called the *commitment*, to Bob.

- Bob chooses a random bit $c$ and sends it to Alice.

    - If $c = 0$, then Alice sends $y = r$ to Bob and Bob checks if the equality $x = y^{-1}t\varphi(y)$ is satisfied.

    - If $c = 1$, then Alice sends $y = sr$ to Bob and Bob checks if the equality $x = y^{-1}w\varphi(y)$ is satisfied.

Again, a correct answer of the prover at the second step leads to acceptance by the verifier.

To break the protocol it is sufficient to find any element $s' \in G$ such that $t = (s')^{-1}w\varphi(s')$ which is an instance of the problem known as the *twisted conjugacy (search) problem*:

Let $G$ be a group. For any $\varphi \in Aut(G)$ and a pair of elements $w, t \in G$ find a *twisted conjugator* for $w$ and $t$, i.e., an element $s \in G$ such that $t = s^{-1}w\varphi(s)$ provided at least one such $s$ exists.

The decision version of this problem is a relatively new algorithmic problem in group theory; it is very nontrivial even for free groups, see [15]. Another class of groups where this problem was considered is the class of polycyclic-by-finite groups [38].

To conclude this section, we mention that recently, two general ways of using Fiat-Shamir's idea were suggested in [56], where it is shown, in particular, that a Fiat-Shamir-like authentication scheme can be arranged so that forgery (a.k.a. impersonation) becomes NP-hard.

## 4.7   Relations between different problems

In this section, we discuss relations between underlying problems in some of the protocols described earlier in this chapter.

We start with the *conjugacy search problem* (CSP), which was used in the protocol described in Section 4.1. A more accurate name for this problem would actually be the *subgroup-restricted* conjugacy search problem:

> Given two elements $w, h$ of a group $G$, a subgroup $A \leq G$, and the information that $w^a = h$ for some $a \in A$, find at least one particular element $a$ like that.

In reference to the Ko-Lee protocol described in Section 4.1, one of the parties (Alice) transmits $w^a$ for some private $a \in A$, and the other party (Bob) transmits $w^b$ for some private $b \in B$, where the subgroups $A$ and $B$ commute elementwise, i.e., $ab = ba$ for any $a \in A$, $b \in B$.

Now suppose the adversary finds $a_1, a_2 \in A$ such that $a_1 w a_2 = a^{-1} w a$ and $b_1, b_2 \in B$ such that $b_1 w b_2 = b^{-1} w b$. Then the adversary gets

$$a_1 b_1 w b_2 a_2 = a_1 b^{-1} w b a_2 = b^{-1} a_1 w a_2 b = b^{-1} a^{-1} w a b = K,$$

the shared secret key.

We emphasize that these $a_1, a_2$ and $b_1, b_2$ do not have to do anything with the private elements originally selected by Alice or Bob, which simplifies the search substantially. We also point out that, in fact, it is sufficient for the adversary to find just one pair, say, $a_1, a_2 \in A$, to get the shared secret key:

$$a_1 (b^{-1} w b) a_2 = b^{-1} a_1 w a_2 b = b^{-1} a^{-1} w a b = K.$$

In summary, to get the secret key $K$, the adversary does not have to solve the (subgroup-restricted) conjugacy search problem, but instead, it is sufficient to solve an apparently easier (subgroup-restricted) decomposition search problem:

> Given two elements $w$ and $w'$ of a group $G$, find any elements $a_1$ and $a_2$ that would belong to a given subgroup $A \leq G$ and satisfy $a_1 \cdot w \cdot a_2 = w'$, provided at least one such pair of elements exists.

We note that the protocol due to Shpilrain and Ushakov considered in Section 4.2.1 is based on a somewhat more general variant of the decomposition search problem where there are two different subgroups:

Given two elements $w$ and $w'$ of a group $G$ and two subgroups $A, B \leq G$, find any two elements $a \in A$ and $b \in B$ that would satisfy $a \cdot w \cdot b = w'$, provided at least one such pair of elements exists.

Then, one more trick reduces the decomposition search problem to a special case where $w = 1$. Namely, given $w' = a \cdot w \cdot b$, multiply it on the left by the element $w^{-1}$ (which is the inverse of the public element $w$) to get

$$w'' = w^{-1}a \cdot w \cdot b = (w^{-1}a \cdot w) \cdot b.$$

Thus, if we denote by $A^w$ the subgroup conjugate to $A$ by the element $w$, the problem for the adversary is now a *factorization search problem*:

Given an element $w'$ of a group $G$ and two subgroups $A^w, B \leq G$, find any two elements $a \in A^w$ and $b \in B$ that would satisfy $a \cdot b = w'$, provided at least one such pair of elements exists.

Since in the original Ko-Lee protocol one has $A = B$, this yields the following interesting observation: if in that protocol $A$ is a normal subgroup of $G$, then $A^w = A$, and the above problem becomes: given $w' \in A$, find any two elements $a_1, a_2 \in A$ such that $w' = a_1a_2$. This problem is trivial: $a_1$ here could be any element from $A$, and then $a_2 = a_1^{-1}w'$.

Therefore, in choosing the platform group $G$ and two commuting subgroups for a protocol described in our Section 4.1 or Section 4.2, one has to avoid normal subgroups. This means, in particular, that "artificially" introducing commuting subgroups as, say, direct factors is inappropriate from the security point of view.

At the other extreme, there are *malnormal* subgroups. A subgroup $A \leq G$ is called malnormal in $G$ if, for any $g \in G$, $A^g \cap A = \{1\}$. We observe that if, in the original Ko-Lee protocol, $A$ is a malnormal subgroup of $G$, then the decomposition search problem corresponding to that protocol has a unique solution if $w \notin A$. Indeed, suppose $w' = a_1 \cdot w \cdot a_1' = a_2 \cdot w \cdot a_2'$, where $a_1 \neq a_2$, say. Then $a_2^{-1}a_1w = wa_2'a_1'^{-1}$, hence $w^{-1}a_2^{-1}a_1w = a_2'a_1'^{-1}$. Since $A$ is malnormal, the element on the left does not belong to $A$, whereas the one on the right does, a contradiction. This argument shows that, in fact, already if $A^w \cap A = \{1\}$ for this particular $w$, then the corresponding decomposition search problem has a unique solution.

Finally, we describe one more trick that reduces, to some extent, the decomposition search problem to the (subgroup-restricted) conjugacy search problem. The same trick reduces the factorization search problem, too, to the subgroup-restricted conjugacy search problem. Suppose we are given $w' = awb$, and we need to recover $a \in A$ and $b \in B$, where $A$ and $B$ are two elementwise commuting subgroups of a group $G$.

Pick any $b_1 \in B$ and compute:

$$[awb, b_1] = b^{-1}w^{-1}a^{-1}b_1^{-1}awbb_1 = b^{-1}w^{-1}b_1^{-1}wbb_1 = (b_1^{-1})^{wb}b_1 = ((b_1^{-1})^w)^b b_1.$$

Since we know $b_1$, we can multiply the result by $b_1^{-1}$ on the right to get $w'' = ((b_1^{-1})^w)^b$. Now the problem becomes: recover $b \in B$ from the known $w'' = ((b_1^{-1})^w)^b$ and $(b_1^{-1})^w$. This is the subgroup-restricted conjugacy search problem. By solving it, one can recover a $b \in B$.

Similarly, to recover an $a \in A$, one picks any $a_1 \in A$ and computes:

$$[(awb)^{-1}, (a_1)^{-1}] = awba_1b^{-1}w^{-1}a^{-1}a_1^{-1} = awa_1w^{-1}a^{-1}a_1^{-1} = (a_1)^{w^{-1}a^{-1}}a_1^{-1}$$
$$= ((a_1)^{w^{-1}})^{a^{-1}}a_1^{-1}.$$

Multiply the result by $a_1$ on the right to get $w'' = ((a_1)^{w^{-1}})^{a^{-1}}$, so that the problem becomes: recover $a \in A$ from the known $w'' = ((a_1)^{w^{-1}})^{a^{-1}}$ and $(a_1)^{w^{-1}}$.

We have to note that, since a solution of the subgroup-restricted conjugacy search problem is not always unique, solving the above two instances of this problem may not necessarily give the right solution of the original decomposition problem. However, any two solutions, call them $b'$ and $b''$, of the first conjugacy search problem differ by an element of the centralizer of $(b_1^{-1})^w$, and this centralizer is unlikely to have a nontrivial intersection with $B$.

A similar computation shows that the same trick reduces the factorization search problem, too, to the subgroup-restricted conjugacy search problem. Suppose we are given $w' = ab$, and we need to recover $a \in A$ and $b \in B$, where $A$ and $B$ are two elementwise commuting subgroups of a group $G$. Pick any $b_1 \in B$ and compute:

$$[ab, b_1] = b^{-1}a^{-1}b_1^{-1}abb_1 = (b_1^{-1})^b b_1.$$

Since we know $b_1$, we can multiply the result by $b_1^{-1}$ on the right to get $w'' = (b_1^{-1})^b$. This is the subgroup-restricted conjugacy search problem. By solving it, one can recover a $b \in B$.

This same trick can, in fact, be used to attack the subgroup-restricted conjugacy search problem itself. Suppose we are given $w' = a^{-1}wa$, and we need to recover $a \in A$. Pick any $b$ from the centralizer of $A$; typically, there is a public subgroup $B$ that commutes with $A$ elementwise; then just pick any $b \in B$. Then compute:

$$[w', b] = [a^{-1}wa, b] = a^{-1}w^{-1}ab^{-1}a^{-1}wab = a^{-1}w^{-1}b^{-1}wab = (b^{-w})^a b.$$

Multiply the result by $b^{-1}$ on the right to get $(b^{-w})^a$, so the problem now is to recover $a \in A$ from $(b^{-w})^a$ and $b^{-w}$. This problem might be easier than the original problem because there is flexibility in choosing $b \in B$. In particular, a feasible attack might be to choose several different $b \in B$ and try to solve the above conjugacy search problem for each in parallel by using some general method (e.g., a length-based attack). Chances are that the attack will be successful for at least one of the $b$s.

# Chapter 5

# Platform Groups

In Section 4.1, we have outlined some of the requirements on the platform group in a protocol based on the conjugacy search problem. Most of these requirements apply, in fact, to platform groups in any "canonical" (i.e., based on a one-way function) cryptographic protocol, so we summarize these general requirements here.

(PG0) The group has to be well known (or well studied, or both).

(PG1) The word problem in $G$ should have a fast (linear- or quadratic-time) solution by a deterministic algorithm. Better yet, there should be an efficiently computable "normal form" for elements of $G$.

(PG2) There should be a way to disguise elements of $G$ so that it would be impossible to recover, say, $x$ and $y$ from a product $xy$ just by inspection. Again, an efficiently computable normal form might be useful here.

In the absence of a normal form, say if $G$ is just given by means of generators and relations without any additional information about properties of $G$, then at least some of these relations should be very short.

(PG3) $G$ should be a group of super-polynomial (i.e., exponential or "intermediate") growth. This means that the number of elements of length $n$ in $G$ should grow faster than any polynomial in $n$; this is needed to prevent attacks by complete exhaustion of the key space. Here "length $n$" is typically just the length of a word representing a group element, but in a more general situation, this could be the length of some other description, i.e., "information complexity".

In this chapter, we are going to examine several groups (or classes of groups) as possible platforms of cryptographic public-key protocols.

## 5.1 Braid groups

Braid groups have appeared as the platform for a "non-commutative" cryptographic public-key protocol in the seminal paper [1]. The legend has it that, after the authors of [1] had invented their protocol, they approached Joan Birman asking her to recommend "good" nonabelian groups that they could use as the platform. Not surprisingly, the answer was "braid groups". After some initial excitement (which has even resulted in naming a new area of "braid group cryptography" — see [25], [31], [41]), it seems now that the conjugacy search problem in a braid group may not provide a sufficient level of security (see [102], [103]), unless keys are selected from some rather narrow subsets (yet to be determined) of the whole group. In what follows, we briefly discuss advantages and disadvantages of braid groups, and then give some group-theoretic background for an interested reader.

It is a fact that abstract groups, unlike numbers, are not something that most people learn at school. There is therefore an obvious communication problem involved in marketing a cryptographic product that uses abstract groups one way or another. Braid groups clearly have an edge here because to explain what they are, one can draw simple pictures, thus alleviating the fear of the unknown. The fact that braid groups cut across many different areas of mathematics (and physics) helps, too, since this gives more credibility to the hardness of the relevant problem (e.g., the conjugacy search problem, see our Section 4.1).

We can recall that, for example, confidence in the security of the RSA cryptosystem is based on literally a centuries-long history of attempts by thousands of people, including such authorities as Euler and Gauss, at factoring integers fast. The history of braid groups goes back to 1927, and again, thousands (well, maybe hundreds) of people, including prominent mathematicians like Artin, Birman, Thurston, V. F. R. Jones, and others have been working on various aspects, including algorithmic ones, of these groups.

On the other hand, from the security point of view, the fact that braid groups cut across so many different areas can be a disadvantage, because different areas provide different tools for solving a problem at hand. Furthermore, braid groups turned out to be linear [8], [85], which makes them potentially vulnerable to linear algebraic attacks (see e.g., [66], [72], [92]), and this alone is a serious security hazard.

As we have said before, braid groups appear in several areas of mathematics, and they admit many equivalent definitions. We start with an explicit presentation by generators and relators.

### 5.1.1   A group of braids and its presentation

In this section we follow the exposition of [36]. A braid is obtained by laying down a number of parallel pieces of string and intertwining them, without losing track of the fact that they run essentially in the same direction. In our pictures the direction is horizontal. We number strands at each horizontal position from the top down. See Figure 5.1 for example.

If we put down two braids $u$ and $v$ in a row so that the end of $u$ matches the beginning of $v$, we get another braid denoted by $uv$, i.e., concatenation of $n$-strand braids is a product. We consider two braids equivalent if there exists an isotopy

Figure 5.1: A 4-strand braid.

between them, i.e., it is possible to move the strands of one of the braids in space (without moving the endpoints of strands and moving strands through each other) to get the other braid. We distinguish a special $n$-strand braid which contains no crossings and call it a trivial braid. Clearly the trivial braid behaves as left and right identity relative to the defined multiplication. The set $B_n$ of isotopy classes of $n$-strand braids has a group structure because, if we concatenate a braid with its mirror image in a vertical plane, the result is isotopic to the trivial braid.

Basically each braid is a sequence of strand crossings. A crossing is called positive if the front strand has a positive slope, otherwise it is called negative. There are exactly $n-1$ crossing types for $n$-strand braids, we denote them by $x_1, \ldots, x_{n-1}$, where $x_i$ is a positive crossing of $i$th and $i+1$st strands. See Figure 5.2 for an example for $B_4$. Since, as we mentioned above, any braid is a sequence

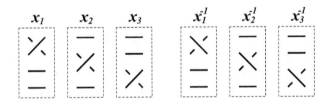

Figure 5.2: Generators of $B_4$ and their inverses.

of crossings the set $\{x_1, \ldots, x_{n-1}\}$ generates $B_n$. It is easy to see that crossings $x_1, \ldots, x_{n-1}$ are subject to the relations

$$[x_i, x_j] = 1$$

for every $i$, $j$ such that $|i - j| > 1$ and

$$x_i x_{i+1} x_i = x_{i+1} x_i x_{i+1}$$

for every $i$ such that $1 \leq i \leq n-2$. The corresponding braid configurations are shown in Figure 5.3. It is more difficult to prove that these two types of relations actually describe the equivalence on braids, i.e., the braid group $B_n$ has the (Artin) presentation

$$B_n = \left\langle \; x_1, \ldots, x_{n-1} \;\middle|\; \begin{matrix} x_i x_j x_i = x_j x_i x_j & \text{if } |i - j| = 1 \\ x_i x_j = x_j x_i & \text{if } |i - j| > 1 \end{matrix} \; \right\rangle.$$

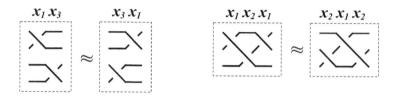

Figure 5.3: Typical relations in braids.

From this description, one easily sees that there are many pairs of commuting subgroups in $B_n$, which makes it possible to use $B_n$ as the platform group for protocols in Sections 4.1 and 4.2. For example, Ko, Lee et al. [84] used the following two commuting subgroups: $LB_n$ generated by $x_1, \ldots, x_{[\frac{n}{2}]-1}$ and $UB_n$ generated by $x_{[\frac{n}{2}]+1}, \ldots, x_{n-1}$.

Now let us go over the properties (PG1)–(PG3) from the introduction to this chapter. There is a vast literature on the word and the conjugacy problems for braid groups. Early solutions of both problems, essentially due to Garside, can be found in J. Birman's monograph [10]. There were many other algorithms suggested since then, most of them quite efficient; the best currently known deterministic algorithm for the word problem has quadratic-time complexity with respect to the length of the input word [36]. We also single out Dehornoy's algorithm for the word problem [29] whose worst case complexity is unknown at the time of this writing, but it appears to work very fast in practical implementations, which suggests that *generic case complexity* of this algorithm might be linear-time. We note, in passing, that there are algorithms with linear-time generic case complexity for the word problem in braid groups, which are based on quotient tests, see [75]. However, these algorithms by design give a fast answer only if the answer is "no", i.e., if the input word is not equal to 1 in $B_n$. Dehornoy's algorithm, on the other hand, gives a fast "yes" answer, too (at least, most of the time). However, no subexponential upper bound for the time complexity of Dehornoy's algorithm has yet been established theoretically.

The conjugacy problem for braid groups was (and is) getting a lot of attention, too. Recently a lot of research was done specifically addressing the (worst case) complexity of this problem; remarkable progress has been made through the work of Birman, Gonzalez-Meneses, Gebhardt, E. Lee, S. J. Lee [11, 13, 12], [40], [46, 47], [91] and others. However, it is still an open problem whether the conjugacy decision and search problems in braid groups can be solved in polynomial time by a deterministic algorithm. A weaker problem, whether the conjugacy problem in braid groups is in the complexity class **NP**, is open, too (see [5, Problem(C3)]).

Getting to the property (PG2), we note that there are several normal forms for elements of braid groups known by now, and these are natural hiding mechanisms. A mild warning here is that the presence of different normal forms might be a potential security hazard because one normal form may reveal what another one is trying to conceal. In fact, this observation was already used in [103] for cryptanalysis of one of the braid-group-based protocols: it was shown that converting Garside normal form to Dehornoy normal form makes the corresponding transmission(s) more vulnerable to some attacks.

Finally, regarding the property (PG3), we note that all braid groups $B_n$ have exponential growth if $n \geq 2$. For $n \geq 3$ this immediately follows from the fact that $B_n$ has free subgroups; for example, $x_1^2$ and $x_2^2$ generate a free subgroup, see [23]. Reasonable estimates of the growth rate of $B_n$ are given in [139].

## 5.1.2 Dehornoy handle free form

Let $w$ be a word in generators of $B_n$. An $x_i$-handle is a subword of $w$ of the form

$$x_i^{-\varepsilon} w(x_1, \dots, x_{i-2}, x_{i+1}, \dots, x_n) x_i^{\varepsilon}$$

where $\varepsilon = \pm 1$. Schematically an $x_i$-handle can be shown as in Figure 5.4. An

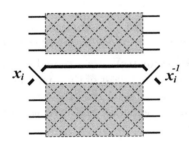

Figure 5.4: A handle.

$x_i$-handle $x_i^{-\varepsilon} w x_i^{\varepsilon}$ where $w = w(x_1, \dots, x_{i-2}, x_{i+1}, \dots, x_n)$ is called *permitted* if $w$ does not contain $x_{i+1}$-handles. We say that the braid word $v'$ is obtained from the braid word $v$ by a *one step handle reduction* if some subword of $v$ is a permitted $x_i$-handle $x_i^{-\varepsilon} w x_i^{\varepsilon}$ and $v'$ is obtained from $v$ by applying the following substitutions for all letters in a handle $x_i^{-\varepsilon} w x_i^{\varepsilon}$:

$$x_j^{\pm 1} \rightarrow \begin{cases} 1 & \text{if } j = i; \\ x_{i+1}^{\varepsilon} x_i^{\pm 1} x_{i+1}^{\varepsilon} & \text{if } j = i+1; \\ x_j^{\pm 1} & \text{if } j < i \text{ or } j > i+1. \end{cases}$$

Schematically a reduction of an $x_i$-handle can be shown as in Figure 5.5. We say that the braid word $v'$ is obtained from the braid word $v$ by $m$ *step handle*

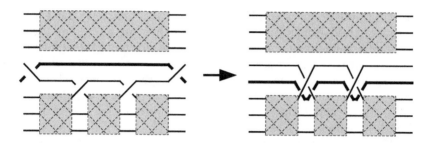

Figure 5.5: One step handle reduction of a permitted $x_i$-handle.

*reduction* if there exists a sequence of $m + 1$ words $v = v_0, v_1, \ldots, v_m = v'$ each of which is obtained from the previous one by a one step handle reduction. A braid word is called *handle free* if it contains no handles. The main statement about handle reduction can be formulated in the following theorem

**Theorem 5.1.1.** *Let $v$ be a braid word. The following holds:*

- *Any sequence of handle reductions applied to $v$ eventually stops and produces a handle free braid word $v'$ (which in general depends on a particular sequence of reductions) representing the same element of the braid group as $v$.*

- *The word $v$ represents identity of a braid group if and only any sequence of handle reductions applied to $v$ produces the trivial word.*

**Remark 5.1.2 (Complexity estimates).** Even though the handle reduction procedure in practice is very efficient and most of the time works in linear time in terms of the length of a braid word there is no good theoretical complexity estimate. For more on strategies for handle reduction and a related discussion on complexity issues see [33, Section 3.3].

### 5.1.3   Garside normal form

Consider a group of permutations $S_n$ on $n$ symbols. With each permutation $s \in S_n$ we can associate the shortest positive braid $\xi_s$ such that $\pi(\xi_s) = s$. The elements

$$S = \{\xi_s \mid s \in S_n\} \subset B_n$$

are called *simple elements*. For permutations $s$ and $t$ we say that a simple element $\xi_s$ is *smaller* than $\xi_t$ (or, that $\xi_s$ is a left divisor of $\xi_t$) and denote it by $\xi_s < \xi_t$ if there exists $r \in S_n$ such that $\xi_t = \xi_s \xi_r$.

There are two special simple braids in $B_n$: the trivial braid which is the smallest element of $S$, and the half-twist braid $\Delta = \xi_{(n,n-1,\ldots,2,1)}$ which is the greatest element of $S$. The set of simple braids with the order $<$ defined above has a lattice structure with gcd and lcm defined by

$$\gcd(\xi_s, \xi_t) = \max \{\xi_r \mid \xi_r < \xi_s \text{ and } \xi_r < \xi_t\}$$

and

$$\operatorname{lcm}(\xi_s, \xi_t) = \min \{\xi_r \mid \xi_s < \xi_r \text{ and } \xi_t < \xi_r\}.$$

Note that since $S$ contains the minimal and the maximal elements it follows that gcd and lcm functions are well defined. Figure 5.6 shows the lattice of simple elements for $B_4$.

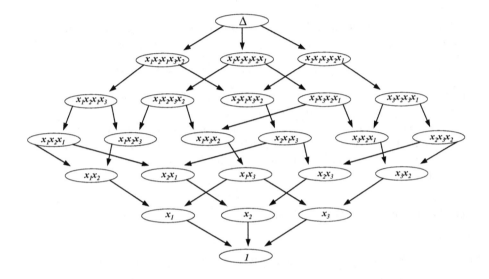

Figure 5.6: The lattice of simple elements in $B_4$.

The *Garside left normal form* of a braid $a \in B_n$ is a pair $(p, (s_1, \ldots, s_l))$ where $p \in \mathbb{Z}$ and $s_1, \ldots, s_l$ is a sequence of permutations in $S_n - \{1, \Delta\}$ satisfying the following property: for each $i = 1, \ldots, l - 1$,

$$\xi_1 = \gcd(\xi_{s_i^{-1}\Delta}, \xi_{s_{i+1}}).$$

A normal form $(p, (s_1, \ldots, s_l))$ represents the following element in $B_n$:

$$\xi_\Delta^p \cdot \xi_{s_1} \cdot \ldots \cdot \xi_{s_n}.$$

**Theorem 5.1.3 (Complexity estimate).** *There exists an algorithm which for any braid word $w = w(x_1, \ldots, x_{n-1})$ computes the normal form of the corresponding braid. Furthermore, the time complexity of the algorithm is $O(n^2|w|^2)$.*

## 5.2 Thompson's group

Thompson's group $F$, like the braid groups $B_n$, is well known in many areas of mathematics, including algebra, geometry, and analysis, so it does satisfy the property (PG0) from the introduction to this chapter. This group, too, is infinite nonabelian.

Now let us briefly go over the properties (PG1)–(PG3). First we note that Thompson's group has the following nice presentation in terms of generators and defining relations:

$$F = \langle x_0, x_1, x_2, \ldots \mid x_i^{-1} x_k x_i = x_{k+1} \ (k > i) \rangle. \tag{5.1}$$

This presentation is infinite. There are also finite presentations of this group; for example,

$$F = \langle x_0, x_1, x_2, x_3, x_4 \mid x_i^{-1} x_k x_i = x_{k+1} \ (k > i, \ k < 4) \rangle,$$

but it is the infinite presentation above that allows for a convenient normal form, so we are going to use that presentation here.

For a survey on various properties of Thompson's group, we refer to [20]. Here we only give a description of the "classical" normal form for elements of $F$.

The classical normal form for an element of Thompson's group is a word of the form

$$x_{i_1} \ldots x_{i_s} x_{j_t}^{-1} \ldots x_{j_1}^{-1}, \tag{5.2}$$

such that the following two conditions are satisfied:

**(NF1)** $i_1 \leq \ldots \leq i_s$ and $j_1 \leq \ldots \leq j_t$

**(NF2)** if both $x_i$ and $x_i^{-1}$ occur, then either $x_{i+1}$ or $x_{i+1}^{-1}$ occurs, too.

We say that a word $w$ is in *seminormal form* if it is of the form (5.2) and satisfies (NF1).

There is an easy procedure for reducing an arbitrary word $w$ to the normal form in Thompson's group. First one reduces the given word to a seminormal form (which is not unique!). This is done by means of the following rewriting system (for all pairs $(i, k)$ such that $i < k$):

$$
\begin{aligned}
x_k x_i &\longrightarrow x_i x_{k+1} \\
x_k^{-1} x_i &\longrightarrow x_i x_{k+1}^{-1} \\
x_i^{-1} x_k &\longrightarrow x_{k+1} x_i^{-1} \\
x_i^{-1} x_k^{-1} &\longrightarrow x_{k+1}^{-1} x_i^{-1}
\end{aligned}
$$

and, in addition, for all $i \in \mathbb{N}$,

$$x_i^{-1} x_i \longrightarrow 1$$

It is fairly obvious that this procedure terminates with a seminormal form for a given word $w$. After a seminormal form is obtained, one has to eliminate "bad pairs", i.e., those pairs $(x_i, x_i^{-1})$ for which the property (NF2) above fails. For this, we need Lemma 5.2.1. For convenience we introduce a parametric function

$$\delta_\varepsilon : \{x_0, x_1, \ldots\}^* \to \{x_0, x_1, \ldots\}^*$$

(where $\varepsilon \in \mathbb{Z}$) defined by

$$x_{i_1} \ldots x_{i_k} \overset{\delta_\varepsilon}{\mapsto} x_{i_1+\varepsilon} \ldots x_{i_k+\varepsilon}.$$

The function $\delta_\varepsilon$ may not be defined for some negative $\varepsilon$, but when it is used, it is assumed that the function is defined.

**Lemma 5.2.1 ([122]).** *Let* $w = x_{i_1} \ldots x_{i_s} x_{j_t}^{-1} \ldots x_{j_1}^{-1}$ *be a seminormal form,* $(x_{i_a}, x_{j_b}^{-1})$ *be the pair of generators in* $w$ *which contradicts* (NF2), *where* $a$ *and* $b$ *are maximal with this property. Let*

$$w' = x_{i_1} \ldots x_{i_{a-1}} \delta_{-1}(x_{i_{a+1}} \ldots x_{i_s} x_{j_t}^{-1} \ldots x_{j_{b+1}}^{-1}) x_{j_{b-1}}^{-1} \ldots x_{j_1}^{-1}.$$

*Then* $w'$ *is in a seminormal form and* $w = w'$ *in* $F$. *Moreover, if* $(x_{i_c}, x_{j_d}^{-1})$ *is the pair of generators in* $w'$ *which contradicts* (NF2) *(where* $a$ *and* $b$ *are maximal with this property), then* $c < a$ *and* $d < b$.

*Proof.* It follows from the definition of (NF2) and seminormal forms that all indices in $x_{i_{a+1}} \ldots x_{i_s} x_{j_t}^{-1} \ldots x_{j_{b+1}}^{-1}$ are greater than $i_a + 1$ and, therefore, indices in $\delta_{-1}(x_{i_{a+1}} \ldots x_{i_s} x_{j_t}^{-1} \ldots x_{j_{b+1}}^{-1})$ are greater than $i_a$. Now it is clear that $w'$ is a seminormal form. Then doing rewrites opposite to those from the rewriting system for obtaining a seminormal form (see above), we can get the word $w'$ from the word $w$. Thus, $w = w'$ in the group $F$.

Now there are two possible cases: either $c > a$ and $d > b$ or $c < a$ and $d < b$. We need to show that the former case is, in fact, impossible. Assume, by way of contradiction, that $c > a$ and $d > b$. Note that if $(x_{i_a}, x_{j_b}^{-1})$ is a pair of generators in $w$ contradicting (NF2), then $(x_{i_a+\varepsilon}, x_{j_b+\varepsilon}^{-1})$ contradicts (NF2) in $\delta_\varepsilon(w)$. Therefore, inequalities $c > a$ and $d > b$ contradict the choice of $a$ and $b$. $\square$

By this lemma, we can start looking for bad pairs in a seminormal form starting in the middle of a word. The next algorithm implements this idea. It is in two parts; the first part finds all bad pairs starting in the middle of a given $w$, and the second part applies $\delta_\varepsilon$ to segments where it is required. A notable feature of Algorithm 5.2.2 is that it does not apply the operator $\delta_{-1}$ immediately (as in $w'$ of Lemma 5.2.1) when a bad pair is found, but instead, it keeps the information about how indices must be changed later. This information is accumulated in two sequences (stacks), one for the positive subword of $w$, the other one for the negative subword of $w$. Also, in Algorithm 5.2.2, the size of stack $S_1$ (or $S_2$) equals the length of an auxiliary word $w_1$ (resp. $w_2$). Therefore, at step B), $x_a$ (resp. $x_b$) is defined if and only if $\varepsilon_1$ (resp. $\varepsilon_2$) is defined.

**Algorithm 5.2.2 (Erasing bad pairs from a seminormal form [122]).**
SIGNATURE. $w = EraseBadPairs(u)$.
INPUT. A seminormal form $u = x_{i_1} \ldots x_{i_s} x_{j_t}^{-1} \ldots x_{j_1}^{-1}$.
OUTPUT. A word $w$ which is the normal form of $u$.
INITIALIZATION. Let $\delta = 0$, $\delta_1 = 0$, $\delta_2 = 0$, $w_1 = 1$, and $w_2 = 1$. Let $u_1 = x_{i_1} \ldots x_{i_s}$

and $u_2 = x_{j_t}^{-1} \ldots x_{j_1}^{-1}$ be the positive and negative parts of $u$. Additionally, we set up two empty stacks $S_1$ and $S_2$.

COMPUTATIONS.

A. Let the current $u_1 = x_{i_1} \ldots x_{i_s}$ and $u_2 = x_{j_t}^{-1} \ldots x_{j_1}^{-1}$.

B. Let $x_a$ be the leftmost letter of $w_1$, $x_b$ the rightmost letter of $w_2$, and $\varepsilon_i$ ($i = 1, 2$) the top element of $S_i$, i.e., the last element that was put there. If any of these values does not exist (because, say, $S_i$ is empty), then the corresponding variable is not defined.

   1) If $s > 0$ and ($t = 0$ or $i_s > j_t$), then:
      a) multiply $w_1$ on the left by $x_{i_s}$ (i.e., $w_1 \leftarrow x_{i_s} w_1$);

      b) erase $x_{i_s}$ from $u_1$;

      c) push 0 into $S_1$;

      d) goto 5).

   2) If $t > 0$ and ($s = 0$ or $j_t > i_s$), then:
      a) multiply $w_2$ on the right by $x_{j_t}^{-1}$ (i.e., $w_2 \leftarrow w_2 x_{j_t}^{-1}$);

      b) erase $x_{j_t}^{-1}$ from $u_2$;

      c) push 0 into $S_2$;

      d) goto 5).

   3) If $i_s = j_t$ and (the numbers $a - \varepsilon_1$ and $b - \varepsilon_2$ (those that are defined) are not equal to $i_s$ or $i_s + 1$), then:
      a) erase $x_{i_s}$ from $u_1$;

      b) erase $x_{j_t}^{-1}$ from $u_2$;

      c) if $S_1$ is not empty, increase the top element of $S_1$;

      d) if $S_2$ is not empty, increase the top element of $S_2$;

      e) goto 5).

   4) If 1)-3) are not applicable (when $i_s = j_t$ and (one of the numbers $a - \varepsilon_1$, $b - \varepsilon_2$ is defined and is equal to either $i_s$ or $i_s + 1$)), then:
      a) multiply $w_1$ on the left by $x_{i_s}$ (i.e., $w_1 \leftarrow x_{i_s} w_1$);

      b) multiply $w_2$ on the right by $x_{j_t}^{-1}$ (i.e., $w_2 \leftarrow w_2 x_{j_t}^{-1}$);

      c) erase $x_{i_s}$ from $u_1$;

      d) erase $x_{j_t}^{-1}$ from $u_2$;

      e) push 0 into $S_1$;

      f) push 0 into $S_2$;

       g) goto 5).

    5) If $u_1$ or $u_2$ is not empty, then goto 1).

C. While $w_1$ is not empty:

    1) let $x_{i_1}$ be the first letter of $w_1$ (i.e., $w_1 = x_{i_1} \cdot w_1'$);

    2) take (pop) $c$ from the top of $S_1$ and add to $\delta_1$ (i.e., $\delta_1 \leftarrow \delta_1 + c$);

    3) multiply $u_1$ on the right by $x_{i_1 - \delta_1}$ (i.e., $u_1 \leftarrow u_1 x_{i_1 - \delta_1}$);

    4) erase $x_{i_1}$ from $w_1$.

D. While $w_2$ is not empty:

    1) let $x_{j_1}^{-1}$ be the last letter of $w_2$ (i.e., $w_2 = w_2' \cdot x_{j_1}^{-1}$);

    2) take (pop) $c$ from the top of $S_2$ and add to $\delta_2$ (i.e., $\delta_2 \leftarrow \delta_2 + c$);

    3) multiply $u_2$ on the left by $x_{j_1 - \delta_2}^{-1}$ (i.e., $u_2 \leftarrow x_{j_1 - \delta_2}^{-1} u_2$);

    4) erase $x_{j_1}^{-1}$ from $w_2$.

E. Return $u_1 u_2$.

We note that in [122], it was shown that the normal form of a given word $w$ in Thompson's group $F$ can be computed in time $O(|w| \log |w|)$.

Finally, we note that recently, Matucci [99] suggested an efficient attack on the decomposition (search) problem in Thompson's group $F$ based on the interpretation of Thompson's group as the group of orientation-preserving piecewise-linear homeomorphisms of the unit interval, where the slopes are powers of 2 and the places where the slope changes are dyadic rationals. The lesson here, as well as with braid groups (see our Section 5.1), is that the presence of different (normal) forms for elements of a group might be a potential security hazard because one (normal) form may reveal what another one is trying to conceal.

## 5.3 Groups of matrices

In this section, we speculate that groups of matrices over finite commutative rings may be the best platforms for "canonical" cryptographic protocols described in our Chapter 4 because these groups have "the best of both worlds" in the sense that matrix multiplication is non-commutative, but matrix entries coming from a commutative ring provides a good hiding mechanism.

Recall that in Section 2.6, we emphasized the need for "diffusion", i.e., for hiding factors in a product, and pointed out that in a group given by generators and relators the only visible mechanism for diffusion is using a normal form for elements of that group. With groups of matrices, we do not have this problem because these groups admit a different description, so that the way we are used to present their elements (as square tables) is, in fact, a "natural" normal form.

The reason why we want the ground ring to be finite is, again, that it is good for diffusion. Finite rings $R$ are *periodic* which means that for any $u \in R$, there are positive integers $m, k$ such that $u^m = u^k$. Periodicity is good for diffusion because it gives rise to a *dynamical system*, and dynamical systems with a large number of states, even "innocent-looking" ones, usually exhibit very complex behavior; it is sufficient to mention the notorious "3x+1" problem [87].

We emphasize once again that

$$COMMUTATIVITY \qquad and \qquad PERIODICITY$$

are two major tools for hiding factors in a product; their importance for cryptographic security in general cannot be overestimated.

At the same time, for better security, commutativity might be reinforced by non-commutativity pretty much the same way as concrete is reinforced by steel to produce ferroconcrete. Thus,

$$COMMUTATIVITY \qquad in\ the\ corset\ of \qquad NON\text{-}COMMUTATIVITY$$

is another important ingredient of cryptographic security. It prevents the attacker from using obvious relations, such as $ab = ba$, to simplify a product.

Finally, we discuss specific finite commutative rings that can be used as ground rings for matrix groups in the cryptographic context. The simplest would be the ring $\mathbb{Z}_n$; matrices over $\mathbb{Z}_n$ can be used as the platform in the original Diffie-Hellman key exchange (see our Section 1.2), but one obvious disadvantage of this ring is that $n$ has to be very large to provide for a sufficiently large key space. We note that there was very little research on how matrix groups over $\mathbb{Z}_n$ compare to $\mathbb{Z}_n$ itself in terms of security when used as the platform for the original Diffie-Hellman key exchange.

Another possibility would be to use $R = \mathbf{F}_p[x]/(f(x))$. Here $\mathbf{F}_p$ is the field with $p$ elements, $\mathbf{F}_p[x]$ is the ring of polynomials over $\mathbf{F}_p$, and $(f(x))$ is the ideal of $\mathbf{F}_p[x]$ generated by an irreducible polynomial $f(x)$ of degree $n$. This ring is actually isomorphic to the field $\mathbf{F}_{p^n}$, but the representation of $R$ as a quotient field allows one to get a large key space while keeping all basic parameters rather small. We note that such a ring has been used by Tillich and Zémor [136] in their construction of a hash function (see also [120]). In the ring they used, $p = 2$, and $n$ is a prime in the range $100 < n < 200$. Matrices that they consider are from the group $SL_2(R)$.

A further improvement of the same idea would be using the ring of *truncated polynomials* over $\mathbb{Z}_n$. Truncated polynomials, or, more precisely, $N$-truncated (one-variable) polynomials, are expressions of the form $\sum_{k=0}^{N} a_k x^k$, with the usual addition and multiplication by the rule $x^i \cdot x^j = x^{(i+j) \bmod (N+1)}$. Just like the ring $R$ in the previous paragraph, this ring is a quotient of an algebra of polynomials by an ideal, but this ideal (generated by the polynomial $x^{N+1}$) is quite simple, and computations in the corresponding quotient ring are quite efficient. Also, a large key space here is provided at a low cost; for example, with $n = 100$, $N = 20$, there are $100^{21} = 10^{42}$ $N$-truncated polynomials. This yields more than $10^{160}$ $2 \times 2$ matrices over this ring.

## 5.4  Small cancellation groups

Small cancellation groups were suggested as platforms in [125].

Small cancellation groups have relators satisfying a simple (and efficiently verifiable) "metric condition" (we follow the exposition in [95]). More specifically, let $F(X)$ be the free group with a basis $X = \{\, x_i \mid i \in I \,\}$, where $I$ is an indexing set. Let $\varepsilon_k \in \{\pm 1\}$, where $1 \leq k \leq n$. A word $w(x_1, \ldots, x_n) = x_{i_1}^{\varepsilon_1} x_{i_2}^{\varepsilon_2} \cdots x_{i_n}^{\varepsilon_n}$ in $F(X)$, with all $x_{i_k}$ not necessarily distinct, is a *reduced X-word* if $x_{i_k}^{\varepsilon_k} \neq x_{i_{k+1}}^{-\varepsilon_{k+1}}$, for $1 \leq k \leq n - 1$. In addition, the word $w(x_1, \ldots, x_n)$ is *cyclically reduced* if it is a reduced $X$-word and $x_{i_1}^{\varepsilon_1} \neq x_{i_n}^{-\varepsilon_n}$. A set $R$ containing cyclically reduced words from $F(X)$ is *symmetrized* if it is closed under cyclic permutations and taking inverses.

Let $G$ be a group with presentation $\langle X; R \rangle$. A nonempty word $u \in F(X)$ is called a *piece* if there are two distinct relators $r_1, r_2 \in R$ of $G$ such that $r_1 = uv_1$ and $r_2 = uv_2$ for some $v_1, v_2 \in F(X)$, with no cancellation between $u$ and $v_1$ or between $u$ and $v_2$. The group $G$ belongs to the class $C(p)$ if no element of $R$ is a product of fewer than $p$ pieces. Also, the group $G$ belongs to the class $C'(\lambda)$ if for every $r \in R$ such that $r = uv$ and $u$ is a piece, one has $|u| < \lambda |r|$.

### 5.4.1  Dehn's algorithm

If $G$ belongs to the class $C'(\frac{1}{6})$, then Dehn's algorithm solves the word problem for $G$ efficiently. This algorithm is very simple:

1. In an input (nonempty) word $w$, look for a "large" piece of a relator from $R$ (that means, a piece whose length is more than a half of the length of the whole relator). If no such piece exists, then output "$w \neq 1$ in $G$".

2. If such a piece, call it $u$, does exist, then $r = uv$ for some $r \in R$, where the length of $v$ is smaller than that of $u$. Then replace $u$ by $v^{-1}$ in $w$. The length of the resulting word $w'$ is smaller than that of $w$. If $w' = 1$, then output "$w = 1$ in $G$".

3. If $w' \neq 1$, then repeat from Step 1 with $w := w'$.

Since the length of $w$ decreases after each loop, this algorithm will terminate in a finite number of steps. It has quadratic time complexity with respect to the length of the input word $w$.

Finally, we note that a generic finitely presented group is a small cancellation group (see [4]).

## 5.5   Solvable groups

Recall that a group $G$ is called *abelian* (or commutative) if $[a, b] = 1$ for any $a, b \in G$, where $[a, b]$ is the notation for $a^{-1}b^{-1}ab$. This can be generalized in different ways. A group $G$ is called *metabelian* if $[[x, y], [z, t]] = 1$ for any $x, y, z, t \in G$. A group $G$ is called *nilpotent of class* $c \geq 1$ if $[y_1, y_2, \ldots, y_{c+1}] = 1$ for any $y_1, y_2, \ldots, y_{c+1} \in G$, where $[y_1, y_2, y_3] = [[y_1, y_2], y_3]$, etc.

The commutator subgroup of $G$ is the group $G' = [G, G]$ generated by all commutators, i.e., by expressions of the form $[u, v] = u^{-1}v^{-1}uv$, where $u, v \in G$. Furthermore, we can define, by induction, the $k$th term of the *lower central series* of $G$: $\gamma_1(G) = G$, $\gamma_2(G) = [G, G]$, $\gamma_k(G) = [\gamma_{k-1}G, G]$. Note that one has $\alpha([u, v]) = [\alpha(u), \alpha(v)]$ for any endomorphism $\alpha$ of $G$. Therefore, $\gamma_k(G)$ is a fully invariant subgroup of $G$ for any $k \geq 1$, and so is $G'' = [G', G']$.

In this section, our focus is on *free metabelian groups* because these groups were used as platforms in a cryptographic protocol in [127].

**Definition 5.5.1.** Let $F_n$ be the free group of rank $n$. The relatively free group $F_n/F_n''$ is called the *free metabelian group* of rank $n$, which we denote by $M_n$.

### 5.5.1   Normal forms in free metabelian groups

In this section, we describe a normal and a seminormal form for elements of a free metabelian group $M_n$. The seminormal form is good for transmissions, because it is easily convertible back to a word representing a transmitted element. However, this form is not unique if $n > 2$ (which is why we call it *seminormal*, so it cannot be used as a shared secret by Alice and Bob in a cryptographic protocol. For the latter purpose, the normal form (a $2 \times 2$ matrix) can be used.

Let $u \in M_n$. By $u_{ab}$ we denote the abelianization of $u$, i.e., the image of $u$ under the natural epimorphism $\alpha : M_n \to M_n/[M_n, M_n]$. Note that we can identify $M_n/[M_n, M_n]$ with $F_n/[F_n, F_n]$. Technically, $u_{ab}$ is an element of a factor group of $F_n$, but we also use the same notation $u_{ab}$ for any word in the generators $x_i$ (i.e., an element of the ambient free group $F_n$) representing $u_{ab}$ when there is no ambiguity.

For $u, v \in M_n$, by $u^v$ we denote the expression $v^{-1}uv$; we also say that $v$ acts on $u$ by conjugation. If $u \in [M_n, M_n]$, then this action can be extended to the group ring $\mathbb{Z}(M_n/[M_n, M_n])$ which we are going to denote by $\mathbb{Z}A_n$, to simplify the notation. (Here $A_n = M_n/[M_n, M_n]$ is the free abelian group of rank $n$.) Let $W \in \mathbb{Z}A_n$ be expressed in the form $W = \sum a_i v_i$, where $a_i \in \mathbb{Z}$, $v_i \in A_n$. Then by $u^W$ we denote the product $\prod (u^{a_i})^{v_i}$. This product is well-defined since any two elements of $[M_n, M_n]$ commute in $M_n$.

Now let $u \in M_n$. Then $u$ can be written in the following seminormal form:

$$u = u_{ab} \cdot \prod_{i<j} [x_i, x_j]^{W_{ij}} \tag{5.3}$$

where $W_{ij} \in \mathbb{Z}A_n$. To get to this form, one can use a "collecting process" based on the following identities (recall that $[x, y] = x^{-1}y^{-1}xy$):

$$[y, x] = [x, y]^{-1},$$
$$xy = yx[x, y],$$
$$xy^{-1} = y^{-1}[y, x]^{y^{-1}x^{-1}}x,$$
$$x^{-1}y = y[y, x]^{y^{-1}x^{-1}}x^{-1},$$
$$[x, y]z = z[x, y]^z.$$

The collecting process itself is simple:

1. Using the above identities, go left to right along the word $u$ collecting all "non-commutator" occurrences of $x_1$ on the left (that means, do not worry about occurrences of the form $[x_1, x_j]$ or $[x_j, x_1]$ created in the process). Repeat this with $x_2$, $x_3$, etc. In the end of this process, $u$ will be written in the form $u = u_{ab} \cdot c$, where $c \in [M_n, M_n]$ is a product of expressions of the form $[x_i, x_j]^g$, $g \in M_n$.

2. Since any two elements of $[M_n, M_n]$ commute in $M_n$, one can now easily re-group the expressions $[x_i, x_j]^g$ so that $u$ takes the form $u = u_{ab} \cdot \prod_{i<j}[x_i, x_j]^{W_{ij}}$, where $W_{ij} \in \mathbb{Z}A_n$.

This process apparently takes quadratic time with respect to the length of $u$.

To convert the seminormal form (5.3) to a word is trivial because (5.3) is, in fact, already a word. The only problem with (5.3) is that it is not unique if $n > 2$, so it cannot be used as a shared secret by Alice and Bob. For the latter purpose, we are now going to introduce a normal form which is unique, efficiently computable (in quadratic time with respect to the length of $u$), but not so easily convertible back to a word.

We have to first introduce *Fox derivatives*, which are noncommutative analogs of usual Leibniz derivatives.

**Definition 5.5.2.** Let $\mathbb{Z}F$ be the group ring of a free group $F$ generated by $x_1, x_2, \ldots$. A *Fox derivation* with respect to $x_i$ is a map $\partial_{x_i} : \mathbb{Z}F \to \mathbb{Z}F$ such that $\partial_{x_i}(x_j) = \delta_{ij}$ and $\partial_{x_i}(vw) = \partial_{x_i}(v) + v \cdot \partial_{x_i}(w)$ for any $v, w \in F$. This map can be extended to the whole $\mathbb{Z}F$ by linearity.

**Example 5.5.3.** Let $g \in F$ and let 1 be the identity of $F$. Since $\partial(1) = \partial(1) + \partial(1)$, it follows that $\partial(1) = 0$. Therefore $\partial(gg^{-1}) = \partial(g) + g\partial(g^{-1}) = 0$, which implies $\partial(g^{-1}) = -g^{-1}\partial(g)$.

**Example 5.5.4.** Let $x$ and $y$ be generators of the free group $F(x, y)$. Then

$$\partial_x([x, y]) = \partial_x(x^{-1}y^{-1}xy)$$
$$= \partial_x(x^{-1}) + x^{-1}\partial_x(y^{-1}) + x^{-1}y^{-1}\partial_x(x) + x^{-1}y^{-1}x\partial_x(y)$$
$$= -x^{-1} + x^{-1}y^{-1} = x^{-1}y^{-1}(1 - y).$$

$$\partial_y([x,y]) = \partial_y(x^{-1}) + x^{-1}\partial_y(y^{-1}) + x^{-1}y^{-1}\partial_y(x) + x^{-1}y^{-1}x\partial_y(y)$$
$$= -x^{-1}y^{-1} + x^{-1}y^{-1}x = -x^{-1}y^{-1}(1-x)\,.$$

Let $F_{ab}$ denote the abelianization of a free group $F$, i.e., the factor group $F/F'$. Let $\alpha : F \to F_{ab}$ be the natural epimorphism; it can be extended to the map $\alpha : \mathbb{Z}F \to \mathbb{Z}F_{ab}$ by linearity. A proof of the following proposition can be found in [57].

**Proposition 5.5.5.** *Let $w \in F_n$. Then $w \in F_n''$ if and only if $\alpha(\partial_{x_i}(w)) = 0$ for each generator $x_i$ of $F_n$.*

This proposition yields a simple algorithm for solving the word problem in a free metabelian group $M_n$: given $w \in M_n$ as a word in relatively free generators $x_i$, one considers $w$ an element of the free group $F_n$ with the same set of free generators, computes $\partial_{x_i}(w)$ for each $x_i$, and checks whether or not all of them abelianize to 0. The latter is straightforward since the word problem in the free abelian group $F_{ab}$ is easily solvable.

This algorithm is not only simple but efficient, too:

**Proposition 5.5.6.** *The algorithm for solving the word problem in $M_n$ based on Proposition 5.5.5 has at most quadratic time complexity with respect to the length of the input word.*

*Proof.* Let $w \in F_n$ and let $|w| = m$ denote the usual lexicographic length of the word $w$. The computation of $\partial_{x_i}(w)$, for any generator $x_i$, produces at most $m$ summands in the free group ring $\mathbb{Z}F_n$. Thus, the computation of $\partial_x$ has at most linear time complexity with respect to $m$. Then, deciding whether or not the abelianization of $\partial_{x_i}(w)$ is 0 amounts to collecting summands of the form $c \cdot h_i$, $c \in \mathbb{Z}, h_i \in F_n$, such that all $h_i$ have the same abelianization. This is achieved by rewriting every $h_i$ in the form $x_1^{a_1}x_2^{a_2} \cdot \ldots \cdot x_n^{a_n} \cdot u_i$, where $u_i \in F_n'$. Since any $h_i$ has length $\leq m$ and the number of $h_i$ is at most $m$, this part of the algorithm takes time $O(m^2)$, which completes the proof.                    □

Finally, we describe the normal form of $u \in M_n$ based on Fox derivatives.

For an element $u \in M_n$ of a free metabelian group, its normal form is a $2 \times 2$ matrix with the following entries:

1. The entry in the lower left corner is 0

2. The entry in the lower right corner is 1

3. The entry in the upper left corner is the abelianization of $u$, so it is an element of the free abelian group $M_n/[M_n, M_n]$

4. The entry in the upper right corner is the most essential one. It is the vector of $n$ abelianized partial Fox derivatives of the word $u$.

We note that the free abelian group $M_n/[M_n, M_n]$ acts on vectors of abelianized Fox derivatives by (componentwise) multiplication. This makes the set of normal forms a group under multiplication. Furthermore, the representation of elements of $M_n$ by their normal forms is faithful, i.e., we actually have an embedding of $M_n$ into a group of matrices; this is called *Magnus embedding* (see e.g., [57]).

## 5.6  Artin groups

Artin groups were used as platforms in a cryptographic protocol in [126].

Let $G(\Gamma)$ be a group with presentation

$$G(\Gamma) = \langle\, g_1, \ldots, g_n \; ; \; r(g_i, g_j) = 1 \ (\text{for } 1 \le i, j \le n \ \text{ and } \ i \ne j)\,\rangle,$$

where $n \ge 2$ and $r(g_i, g_j) = 1$ is a relator involving two generators. Given $G(\Gamma)$ there is an associated labeled graph $\Gamma_G$ and vice versa. The vertices of the graph $\Gamma_G$ are labeled by the generators of $G(\Gamma)$. Any two vertices $g_i, g_j \in \Gamma_G$ are connected by an edge if there is a relation $r(g_i, g_j) \in G$ between the corresponding generators; in other words, edges are labeled by relations.

**Example 5.6.1.** An *Artin group* $A(\Gamma)$ is a group with presentation

$$A(\Gamma) = \langle\, a_1, \ldots, a_n \; ; \; \mu_{ij} = \mu_{ji} \text{ for } 1 \le i < j \le n)\,\rangle, \quad \text{where } \mu_{ij} = \underbrace{a_i\, a_j\, a_i \ldots}_{m_{ij}}$$

and $m_{ij} = m_{ji}$. Artin groups arise as generalizations of braid groups, see e.g., [3]. For an Artin group $A(\Gamma)$, the associated labeled graph $\Gamma_A$ has no multiple edges or loops. The vertices $a_i$ of $\Gamma_A$ are the generators of the Artin group. Any two vertices $a_i, a_j \in \Gamma_A$ are connected by an edge, labeled with the integer $m_{ij}$, associated to the relation $\mu_{ij} = \mu_{ji}$ (between the corresponding generators $a_i, a_j \in A(\Gamma)$).

In general, automorphisms (or endomorphisms) of the graph $\Gamma_G$ induce automorphisms (or endomorphisms) of the group $G(\Gamma)$. Therefore, the graph associated to $G(\Gamma)$ gives us a way to construct a semigroup of endomorphisms of $G(\Gamma)$ that can contain a large pool of commuting elements. This was used in [126] as a basis of a key exchange protocol.

**Example 5.6.2.** The relations of the braid groups $B_n$ involve two generators. The corresponding graph associated to $B_n$ is just a simple path, and it has only one automorphism that induces the following automorphism of $B_n$: $\sigma_i \mapsto \sigma_{n-i}$, which happens to be an inner automorphism of $B_n$. For other $G(\Gamma)$ groups, however, their corresponding graphs are more complex, and it is easy to arrange for a large semigroup (or a group) $T \subseteq End\, G(\Gamma)$ of endomorphisms (or automorphisms).

Artin groups $A(\Gamma)$ with the property that all the integers $m_{ij} \ge 4$ are called *Artin groups of extra large type*. A tree $\Gamma_A$ can be associated to an Artin group of extra large type, providing a direct procedure for constructing a semigroup of endomorphisms of $A(\Gamma)$. Moreover, Artin groups of extra large type are automatic

[116], thus the word problem for groups in this class can be solved in quadratic time, and by a result of [75], the word problem is solvable generically in linear time.

## 5.7   Grigorchuk's group

In this section we define the original Grigorchuk group $\Gamma$ which first appeared in [52]. The group $\Gamma$ is very important for many group theoretic problems, such as growth [53], amenability [54], just finite groups [55]. Recently it was also considered as a possible platform for a cryptographic scheme [117]. In our exposition we follow the book [63].

We start out by discussing the group of automorphisms of the infinite rooted binary tree. Denote by $\mathcal{T}$ the infinite rooted binary tree. The set $T$ of vertices of $\mathcal{T}$ is the set of finite binary sequences as shown in Figure 5.7. Sets of vertices

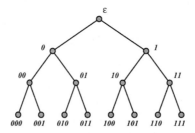

Figure 5.7: Binary tree with labeled vertices.

$L_k = \{b_1 \ldots b_k \mid b_i = 0, 1\}$ of the same length $k$ are called *levels* in the tree $\mathcal{T}$. An automorphism $\varphi$ of $\mathcal{T}$ fixes the root $\varepsilon$ of $\mathcal{T}$ and permutes the vertices preserving the connectedness, i.e., if vertices $v_1$ and $v_2$ are connected by an edge in $\mathcal{T}$, then so are $\varphi(v_1)$ and $\varphi(v_1)$. It is easy to see that for any $k$ an automorphism $\varphi$ of $\mathcal{T}$ defines a permutation of the vertices of the level $L_k$. The group of automorphisms of $\mathcal{T}$ is denoted by $Aut(\mathcal{T})$.

Consider an automorphism $a$ which plays a special role in $Aut(\mathcal{T})$. It maps a $\{0,1\}$-sequence to a $\{0,1\}$-sequence obtained by changing the first element:

$$a(b_1, b_2, \ldots, b_n) = (1 - b_1, b_2, \ldots, b_n).$$

Geometrically the action of $a$ can be seen as swapping the left and the right subtrees of $\mathcal{T}$. Denote a subgroup of $Aut(\mathcal{T})$ stabilizing the vertices $L_1$ by $St(1)$. The subgroup $St(1)$ is normal of index 2 in $Aut(\mathcal{T})$ with left (and right) cosets $\{1, a\}$.

The Grigorchuk group $\Gamma$ is a subgroup of $Aut(\mathcal{T})$ generated by four elements traditionally denoted by $a, b, c, d$. The element $a$ is defined above. The other three automorphisms are defined as follows:

$$\mathbf{b}(b_1, b_2, \ldots, b_n) = \begin{cases} (b_1, 1 - b_2, \ldots, b_n), & \text{if } b_1 = 0; \\ (b_1, \mathbf{c}(b_2, \ldots, b_n)), & \text{if } b_1 = 1; \end{cases}$$

$$\mathbf{c}(b_1, b_2, \ldots, b_n) = \begin{cases} (b_1, 1 - b_2, \ldots, b_n), & \text{if } b_1 = 0; \\ (b_1, \mathbf{d}(b_2, \ldots, b_n)), & \text{if } b_1 = 1; \end{cases}$$

$$\mathbf{d}(b_1, b_2, \ldots, b_n) = \begin{cases} (b_1, b_2, \ldots, b_n), & \text{if } b_1 = 0; \\ (b_1, \mathbf{b}(b_2, \ldots, b_n)), & \text{if } b_1 = 1; \end{cases}$$

It is straightforward to check that so defined maps belong to $Aut(\mathcal{T})$ and that $b, c, d \in St(1)$. The next example demonstrates the computation of the image of a vertex (binary sequence):

$$b(1, 1, 1, 0, 1) = (1, c(1, 1, 0, 1)) = (1, 1, d(1, 0, 1))$$
$$= (1, 1, 1, b(0, 1)) = (1, 1, 1, 0, 0).$$

**Lemma 5.7.1.** *The relations*

$$a^2 = b^2 = c^2 = d^2 = 1 \quad and \quad bc = cb = d$$

*hold for generators of* $\Gamma$.

*Proof.* Straightforward to check. $\qquad\square$

**Lemma 5.7.2.** *The group* $\Gamma = \langle a, b, c, d \rangle$ *is a quotient of* $(\mathbb{Z}/2\mathbb{Z}) * ((\mathbb{Z}/2\mathbb{Z}) \times (\mathbb{Z}/2\mathbb{Z}))$.

*Proof.* It follows from Lemma 5.7.1 that a subgroup $\langle b, c, d \rangle$ is isomorphic to $(\mathbb{Z}/2\mathbb{Z}) \times (\mathbb{Z}/2\mathbb{Z})$. Furthermore, we have a relation $a^2 = 1$. Thus a mapping from $(\mathbb{Z}/2\mathbb{Z}) * ((\mathbb{Z}/2\mathbb{Z}) \times (\mathbb{Z}/2\mathbb{Z}))$ to $\Gamma$ which maps the generator of the first free factor to $a$ and abelian generators of the second free factor to $b$ and $c$ is an epimorphism. $\quad\square$

It follows from Lemma 5.7.1 that any word $w = w(a, b, c, d)$ is equal to a word of the type

$$u_0 a u_1 a u_2 \ldots u_{k-1} a u_k \tag{5.4}$$

where $u_0, \ldots, u_k \in \{b, c, d\}$ and, perhaps, $u_0$, $u_k$ are trivial. A word of the type (5.4) is called *reduced*.

**Lemma 5.7.3.** *A word* $w(a, b, c, d)$ *represents an element of* $St(1)$ *if and only if the total number of a symbols is even. Furthermore, the subgroup* $St(1)$ *of* $\Gamma$ *is generated by six elements* $\{b, c, d, aba, aca, ada\}$.

*Proof.* Follows from the definition of the elements $a$, $b$, $c$, and $d$. $\qquad\square$

It is straightforward to represent a word representing an element in $St(1)$ as a product of the generators of $St(1)$ of the form

$$u_0 \cdot (au_1 a) \cdot u_2 \cdot (au_3 a) \cdot u_4 \ldots \cdot (au_{k-3} a) \cdot u_{k-2} \cdot (au_{k-1} a) \cdot u_k \tag{5.5}$$

where $u_0, \ldots, u_k \in \{b, c, d\}$ and $u_0$, $u_k$ are, perhaps, trivial. Since any element $\varphi$ of $St(1)$ acts trivially on the first level of the tree $\mathcal{T}$ it follows that $\varphi$ can be represented as a pair $(\varphi_0, \varphi_1)$ where $\varphi_0$ and $\varphi_1$ are automorphisms of the left and the right subtrees of $\mathcal{T}$ respectively defined by $\varphi$. It is easy to check that

$$
\begin{aligned}
b &= (a, c), & aba &= (c, a), \\
c &= (a, d), & aca &= (d, a), \\
d &= (1, b), & ada &= (b, 1).
\end{aligned}
\tag{5.6}
$$

Moreover, if $u, v \in St(1)$ and $u = (u_0, u_1)$ and $v = (v_0, v_1)$, then

$$
uv = (u_0 v_0, u_1 v_1).
\tag{5.7}
$$

**Lemma 5.7.4.** *Let* $w = w(a, b, c, d) \in St(1)$ *be a reduced word and* $w = (w_0, w_1)$. *Then*

$$
|w_0|, |w_1| \leq \frac{|w| + 1}{2}.
$$

*Moreover, if* $w$ *starts with a symbol* $a$, *then* $|w_0|, |w_1| \leq \frac{|w|}{2}$.

*Proof.* Any word $w \in St(1)$ can be seen as a product (5.5). Using formulae (5.6) and (5.7) it is trivial to check both inequalities. $\qquad \square$

**Algorithm 5.7.5 (Identity Problem for $\Gamma$).**
INPUT. A reduced word $w = w(a, b, c, d)$.
OUTPUT. *True* if $w$ represents the identity of $\Gamma$. *False* otherwise.
COMPUTATIONS.

A. If $|w| = 0$, then output *True*. If $|w| = 1$, then output *False*.

B. Cyclically permute $w$ so it starts with the $a$ symbol.

C. Compute the algebraic sum of powers of $a$ in $w$. If it is odd, then output *False*.

D. Rewrite $w$ as a product of elements $b, c, d, aba, aca, ada$ and using (5.6) find corresponding $(w_0, w_1)$.

E. Recursively check if $w_0$ and $w_1$ represent identity of $\Gamma$ and if so output *True*. Otherwise output *False*.

**Theorem 5.7.6.** *Algorithm 5.7.5 solves the Identity problem for the Grigorchuk group* $\Gamma$ *in time* $O(|w| \log |w|)$.

*Proof.* A word $w$ represents the identity in $\Gamma$ if and only if it belongs to $St(1)$ and acts trivially on the left and the right subtree of $\mathcal{T}$. Hence, the algorithm for solving the identity problem must start with checking if $w \in St(1)$ and if so rewrite $w$ as a product of elements $b, c, d, aba, aca, ada$. Then using the equalities (5.6) compute elements $(w_0, w_1)$ describing the action on subtrees and check if they represent identities. Thus Algorithm 5.7.5 correctly solves the Identity problem for $\Gamma$.

For a word $w$, Algorithm 5.7.5 performs Steps A–D in linear time $O(|w|)$. At Step E we come up to two words $w_0$ and $w_1$ each of length less than $\frac{|w|}{2}$ and recursively run the same procedure for $w_0$ and $w_1$.

The total time complexity of steps A–D for words $w_0$ and $w_1$ is $O(|w_0|) + O(|w_1|) = O(|w|)$ and on step D we split $w_0$ and $w_1$ and obtain four new words. The time complexity of steps A–D for those words is bounded by $O(|w|)$ again. And so on until we get words of lengths 0 and 1 for which we know the answer. Since each time we split words into two pieces each of which is not longer than the half of the original word, it follows that in $\log_2 |w|$ steps we come up to words of lengths 0 or 1. Therefore the total complexity of the process is bounded by $O(|w| \log_2 |w|)$. $\qquad\square$

Based on Algorithm 5.7.5 one can arrange a procedure for constructing normal forms of elements of $\Gamma$. The normal form $\rho_w$ of an element $w \in \Gamma$ is a finite binary tree with vertices labeled with elements $\{1, a, b, c, d\} \in \Gamma$. The tree $\rho_w$ for

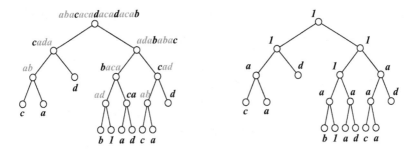

Figure 5.8: A normal form of the element $abacacadacadacab$ (on the right). The tree on the left visualizes the process of computing the normal forms.

a word $w = w(a, b, c, d)$ can be constructed recursively as follows:

- Using Algorithm 5.7.5 check if $w$ is equal to $1, a, b, c, d$ in $\Gamma$ and if so $\rho_w$ contains one vertex labeled with the corresponding symbol.

- If $w$ is not equal to $1, a, b, c, d$ in $\Gamma$, then:

    - If $w \in St(1)$ and $w = (w_0, w_1)$, then $\rho_w$ is a tree with a root labeled by 1, the left subtree is $\rho_{w_0}$, and the right subtree is $\rho_{w_1}$.

    - If $w \notin St(1)$ and $wa = (w_0, w_1)$, then $\rho_w$ is a tree with a root labeled by $a$, the left subtree is $\rho_{w_0}$, and the right subtree is $\rho_{w_1}$.

We omit proofs for the next two statements.

**Proposition 5.7.7.** Let $u = u(a, b, c, d)$ and $v = v(a, b, c, d)$. The words $u$ and $v$ represent the same element of $\Gamma$ if and only if $\rho_u = \rho_v$.

# Chapter 6

# Using Decision Problems in Public Key Cryptography

In this chapter, we suggest using *decision problems* from combinatorial group theory as the core of a public key establishment protocol or a public key cryptosystem. Decision problems are problems of the following nature: given a property $P$ and an object $O$, find out whether or not the object $O$ has the property $P$. Decision problems may allow us to address (to some extent) the following challenge of public key cryptography: to design a cryptosystem that would be secure against (at least, some) "brute force" attacks by an adversary with essentially unlimited computational capabilities.

A particular decision problem that we consider here is the *word problem* which is: given a recursive presentation of a group $G$ and an element $g \in G$, find out whether or not $g = 1$ in $G$. From the very description of the word problem we see that it consists of two parts: "whether" and "not". We call them the "yes" and "no" parts of the word problem, respectively. If a group is given by a recursive presentation in terms of generators and relators, then the "yes" part of the word problem has a recursive solution because one can recursively enumerate all products of defining relators, their inverses and conjugates. However, the number of factors in such a product required to represent a word of length $n$ which is equal to 1 in $G$, can be very large compared to $n$; in particular, there are groups $G$ with efficiently solvable word problem and words $w$ of length $n$ equal to 1 in $G$, such that the number of factors in any factorization of $w$ into a product of defining relators, their inverses and conjugates is not bounded by any tower of exponents in $n$, see [117]. Furthermore, if in a group $G$ the word problem is recursively unsolvable, then the length of a proof verifying that $w = 1$ in $G$ is not bounded by any recursive function of the length of $w$.

We also note that the "no" part of the word problem in many groups is recursively unsolvable, and therefore the "brute force" attack described above will not be effective against this part. We have to point out though that there is no recursively presented group (or semigroup) that would have both "yes" and "no" parts of the word problem recursively unsolvable.

# 6.1   The Shpilrain-Zapata scheme

In this section, we describe a cryptographic protocol (see [125]) that employs computational hardness of the word problem in groups. In this protocol, Bob transmits to Alice an encrypted binary sequence which Alice decrypts correctly with probability "very close" to 1.

We note that a long time ago, there was an attempt to use the word problem in public key cryptography [97], but it did not meet with success, for several reasons. One of the reasons, which is relevant to the discussion above, was pointed out recently in [5]: the problem which is actually used in [97] is not the word problem, but the *word choice problem*: given $g, w_1, w_2 \in G$, find out whether $g = w_1$ or $g = w_2$ in $G$, provided one of the two equalities holds. In this problem, both parts are recursively solvable for any recursively presented platform group $G$ because they both are the "yes" parts of the word problem, and therefore the word choice problem cannot be used for our purposes. Thus, a similarity of the proposal of [125] to that of [97] is misleading, and the former seems to be the first proposal actually based on a decision problem.

## 6.1.1   The protocol

Here is a sketch of the protocol from [125]; details are given in the following sections.

**Protocol:**

1. A pool of group presentations with efficiently solvable word problem is considered public (e.g., is part of Alice's software).

2. Alice chooses randomly a particular presentation $\Gamma$ from the pool, diffuses it by isomorphism-preserving transformations to obtain a diffused presentation $\Gamma'$, discards some of the relators and publishes the abridged diffused presentation $\hat{\Gamma}$.

3. Bob transmits his private binary sequence to Alice by transmitting an element equal to 1 in $\hat{\Gamma}$ (and therefore also in $\Gamma'$) in place of "1" and an element not equal to 1 in $\Gamma'$ in place of "0".

4. Alice recovers Bob's binary sequence by first converting elements of $\Gamma'$ to the corresponding (under the isomorphism that she knows) elements of $\Gamma$, and then solving the word problem in $\Gamma$.

Most parts of this protocol are rather nontrivial and open several interesting research avenues. We discuss parts 1., 2., 3. in our Sections 6.1.2, 6.1.3, 6.1.4, respectively.

*A priori* it looks like the most nontrivial part is finding an element which is not equal to 1 in $\Gamma'$ since Bob does not even know the whole presentation $\Gamma'$. We solve this problem by "going with the flow", so to speak. More specifically, we just let Bob select a random (well, almost random) word of sufficiently big length and show that, with overwhelming probability, such an element is not equal to 1 in $\Gamma'$. We discuss this in more detail in Section 6.1.4.

We emphasize once again what is, in our opinion, the main novelty of this protocol compared to the existing ones. The point is to deprive the adversary (Eve) from attacking the protocol by doing an exhaustive search of the sender's private key space, which is the most obvious (although often "computationally infeasible") way to attack all existing public key protocol.

The way we plan to achieve our goal is relevant to part 3. of the above protocol, more specifically, to solving the word problem in $\hat{\Gamma}$. If Bob transmits an element $g$ equal to 1 in $\hat{\Gamma}$, Eve may be able to detect this by going over all products of all conjugates of relators from $\hat{\Gamma}$ and their inverses. This set is recursive, but as we have pointed out in the Introduction, there are groups $G$ with efficiently solvable word problem and words $w$ of length $n$ equal to 1 in $G$, such that the length of a proof verifying that $w = 1$ in $G$ is not bounded by any tower of exponents in $n$, see [117].

Furthermore, if Bob transmits an element $g$ not equal to 1 in $\hat{\Gamma}$, then detecting this is even more difficult for Eve. In fact, it is impossible in general; Eve's best hope here is that she will be lucky enough to find a factor group of $\hat{\Gamma}$ where the word problem is solvable, and that $g \neq 1$ in that factor group. This is what we call a *quotient attack*, see our Section 6.1.6.

Now let us take a closer look at Eve's "encryption emulation" attack, which is:

> Eve performs key generations over and over again, each time with fresh randomness, until the transmission to be attacked is obtained – this will happen eventually with overwhelming probability. The correctness of the scheme guarantees that the corresponding secret key (as obtained by Eve while performing key generation) allows her to decrypt illegitimately.

This would indeed be viable if the correctness of the legitimate decryption by Alice was perfect. However, in our situation this kind of attack may not work for a general $\hat{\Gamma}$. Suppose Eve is building up two lists, corresponding to two possible encryptions " $0 \to w \neq 1$ in $\hat{\Gamma}$" or "$1 \to w = 1$ in $\hat{\Gamma}$" by Bob. Our first observation is that the list that corresponds to " $0 \to w \neq 1$" is useless to Eve because it is simply going to contain all words in the alphabet $X = \{x_1, \ldots, x_n, x_1^{-1}, \ldots, x_n^{-1}\}$ since Bob is choosing such $w$ simply as a random word. Therefore, Eve may just as well forget about this list and concentrate on the other one, which corresponds to "$1 \to w = 1$".

Now the situation boils down to the following: if a word $w$ transmitted by Bob appears on the list, then it is equal to 1 in $G$. If not, then not. The only problem is: how can Eve possibly conclude that $w$ does not appear on the list if the list is infinite? One could say here that Eve can stop at some point and conclude that $w \neq 1$ with overwhelming probability, just like Alice does. The point however is that this probability may not at all be as "overwhelming" as the probability of the correct decryption by Alice. Compare:

1. For Alice to decrypt correctly "with overwhelming probability", the probability $P_1(N)$ for a random word $w$ of length $N$ not to be equal to 1 should converge to 1 (reasonably fast) as $N$ goes to infinity.

2. For Eve to decrypt correctly "with overwhelming probability", the probability $P_2(N, f(N))$ for a random word $w$ of length $N$, which is equal to 1, to have a proof of length $\leq f(N)$ verifying that $w = 1$, should converge to 1 (reasonably fast) as $N$ goes to infinity. Here $f(N)$ represents Eve's computational capabilities; this function can be arbitrary, but fixed.

We see that the functions $P_1(N)$ and $P_2(N, f(N))$ are of very different nature, and any correlation between them is unlikely. We note that the function $P_1(N)$ is generally well understood, and in particular, it is known that in any infinite group $G$, $P_1(N)$ indeed converges to 1 as $N$ goes to infinity; see our Section 6.1.4 for more details.

On the other hand, the functions $P_2(N, f(N))$ are more complex; they are currently subject of very active research, and in particular, it may happen that for any $f(N)$, there are groups in which $P_2(N, f(N))$ does not converge to 1 at all. Of course, $P_2(N, f(N))$ may depend on a particular algorithm used by Bob to produce words equal to 1, but we leave this discussion to another occasion.

To conclude this section, we point out that encryption (of one bit) in this protocol is rather efficient; it takes quadratic time in the length of a transmitted word; the latter is approximately 150 on average, according to our computer experiments. Also, it is straightforward to see that the time Alice needs to decrypt each transmitted word $w$ is bounded by $C \cdot |w|$, where $|w|$ is the length of $w$ and $C$ is a constant which basically depends on Alice's private isomorphism between $\Gamma$ and $\Gamma'$.

The fact that Alice (the receiver) and the adversary are separated in power is essentially due to Alice's knowledge of her private isomorphism between $\Gamma$ and $\Gamma'$ (note that Bob does *not* have to know this isomorphism for encryption!).

We have to admit here one disadvantage of this protocol compared to most well-established public key protocols: we have encryption with a rather big "expansion factor". Computer experiments show that, with suggested parameters, one bit in Bob's message gets encrypted into a word of length approximately 150 on average. This is the price we have to pay for granting the adversary too much computational power.

Finally, we touch upon semantic security in the end of Section 6.1.4.

## 6.1.2 Pool of group presentations

There are many classes of finitely presented groups with solvable word problem known by now, e.g., one-relator groups, hyperbolic groups, nilpotent groups, metabelian groups. Note however that Alice should be able to randomly select a presentation from the pool *efficiently*, which imposes some restrictions on classes of presentations that can be used in this context. The class of finitely presented groups that we suggest including in our pool is the class of *small cancellation groups*, see Section 5.4 of this book.

We briefly recall some basic definitions here. Small cancellation groups have relators satisfying a simple (and efficiently verifiable) "metric condition" . More specifically, let $F(X)$ be the free group with a basis $X = \{ x_i \,|\, i \in I \}$, where $I$ is an indexing set. Let $\varepsilon_k \in \{\pm 1\}$, where $1 \le k \le n$. A word $w(x_1, \ldots, x_n) = x_{i_1}^{\varepsilon_1} x_{i_2}^{\varepsilon_2} \cdots x_{i_n}^{\varepsilon_n}$ in $F(X)$, with all $x_{i_k}$ not necessarily distinct, is a *reduced X-word* if $x_{i_k}^{\varepsilon_k} \neq x_{i_{k+1}}^{-\varepsilon_{k+1}}$, for $1 \le k \le n-1$. In addition, the word $w(x_1, \ldots, x_n)$ is *cyclically reduced* if it is a reduced $X$-word and $x_{i_1}^{\varepsilon_1} \neq x_{i_n}^{-\varepsilon_n}$. A set $R$ containing cyclically reduced words from $F(X)$ is *symmetrized* if it is closed under cyclic permutations and taking inverses.

Let $G$ be a group with presentation $\langle X; R \rangle$. A nonempty word $u \in F(X)$ is called a *piece* if there are two distinct relators $r_1, r_2 \in R$ of $G$ such that $r_1 = uv_1$ and $r_2 = uv_2$ for some $v_1, v_2 \in F(X)$, with no cancellation between $u$ and $v_1$ or between $u$ and $v_2$. The group $G$ belongs to the class $C(p)$ if no element of $R$ is a product of fewer than $p$ pieces. Also, the group $G$ belongs to the class $C'(\lambda)$ if for every $r \in R$ such that $r = uv$ and $u$ is a piece, one has $|u| < \lambda |r|$.

In particular, if $G$ belongs to the class $C'(\frac{1}{6})$, then Dehn's algorithm solves the word problem for $G$ efficiently (see Section 5.4). This algorithm has quadratic time complexity with respect to the length of the input word $w$.

We also note that a generic finitely presented group is a small cancellation group (see [4]); therefore, to randomly select a small cancellation group, Alice can just take a few random words and check whether the corresponding symmetrized set satisfies the condition for $C'(\frac{1}{6})$. If not, then repeat.

To conclude this section, we give a more specific recipe, with sample parameters, for Alice to produce a presentation $\Gamma$ for the protocol in Section 6.1.1.

1. Alice fixes a number $k$, $10 \le k \le 20$, of generators in her presentation $\Gamma$. Her $\Gamma$ will therefore have generators $x_1, \ldots, x_k$.

2. Alice selects $m$ random words $r_1, \ldots, r_m$ in the generators $x_1, \ldots, x_k$. Here $10 \le m \le 30$ and the lengths $l_i$ of $r_i$ are random integers from the interval $L_1 \le l_i \le L_2$. Particular values that we suggest are: $L_1 = 12$, $L_2 = 20$.

3. After Alice obtains the abridged presentation $\hat{\Gamma}$, she adds a relation

$$x_i' = \prod_{j=1}^{M} [x_i', w_j]$$

to it, where $x_i'$ is a (randomly chosen) generator from $\hat{\Gamma}$, $w_j$ are random elements of length 1 or 2 in the generators $x_1', x_2', \ldots$, and $M = 10$. (Our commutator notation is: $[a, b] = a^{-1}b^{-1}ab$.) This relation is needed to foil *quotient attacks*, see Section 6.1.6. Then Alice finds the preimage of this relation under the isomorphism between $\Gamma$ and $\hat{\Gamma}$ and adds this preimage to the defining relators of $\Gamma$. Thus, $\Gamma$ finally has $k$ generators and $m+1$ defining relators.

4. Finally, Alice checks whether her private presentation $\Gamma$ satisfies the small cancellation condition $C'(\frac{1}{6})$ (it will with overwhelming probability, see [4]). If not, then she has to start over.

## 6.1.3   Tietze transformations: elementary isomorphisms

In this section, we explain how Alice can implement step (2) of the protocol given in Section 6.1.1. First we introduce Tietze transformations (see also our Section 2.5); these are "elementary isomorphisms": any isomorphism between finitely presented groups is a composition of Tietze transformations. What is important to us is that every Tietze transformation is easily invertible, and therefore Alice can compute the inverse isomorphism that takes $\Gamma'$ to $\Gamma$.

Tietze introduced isomorphism-preserving elementary transformations that can be applied to groups presented by generators and relators. They are of the following types.

**(T1)** *Introducing a new generator*: Replace $\langle x_1, x_2, \ldots \mid r_1, r_2, \ldots \rangle$ by $\langle y, x_1, x_2, \ldots \mid ys^{-1}, r_1, r_2, \ldots \rangle$, where $s = s(x_1, x_2, \ldots)$ is an arbitrary element in the generators $x_1, x_2, \ldots$.

**(T2)** *Canceling a generator* (this is the converse of (T1)): If we have a presentation of the form $\langle y, x_1, x_2, \ldots \mid q, r_1, r_2, \ldots \rangle$, where $q$ is of the form $ys^{-1}$, and $s, r_1, r_2, \ldots$ are in the group generated by $x_1, x_2, \ldots$, replace this presentation by $\langle x_1, x_2, \ldots \mid r_1, r_2, \ldots \rangle$.

**(T3)** *Applying an automorphism*: Apply an automorphism of the free group generated by $x_1, x_2, \ldots$ to all the relators $r_1, r_2, \ldots$.

**(T4)** *Changing defining relators*: Replace the set $r_1, r_2, \ldots$ of defining relators by another set $r_1', r_2', \ldots$ with the same normal closure. That means, each of $r_1', r_2', \ldots$ should belong to the normal subgroup generated by $r_1, r_2, \ldots$, and vice versa.

Tietze proved that two groups $\langle x_1, x_2, \ldots \mid r_1, r_2, \ldots \rangle$ and $\langle x_1, x_2, \ldots \mid s_1, s_2, \ldots \rangle$ are isomorphic if and only if one can get from one of the presentations to the other by a sequence of transformations (T1)–(T4).

For each Tietze transformation of the types (T1)–(T3), it is easy to obtain an explicit isomorphism (as a mapping on generators) and its inverse. For a Tietze transformation of the type (T4), the isomorphism is just the identity map. We would like here to make Tietze transformations of the type (T4) recursive, because *a priori* it is not clear how Alice can actually apply these transformations. Thus, Alice is going to use the following recursive version of (T4):

**(T4′)** In the set $r_1, r_2, \ldots$, replace some $r_i$ by one of the: $r_i^{-1}$, $r_i r_j$, $r_i r_j^{-1}$, $r_j r_i$, $r_j r_i^{-1}$, $x_k^{-1} r_i x_k$, $x_k r_i x_k^{-1}$, where $j \neq i$, and $k$ is arbitrary.

We suggest that in part 2. of the protocol in Section 6.1.1, Alice should first apply several transformations of the type (T4′) to "mix" the presentation $\Gamma$. (This does not add complexity to the final isomorphism since for a Tietze transformation of the type (T4), the isomorphism is just the identity map, as we have noted above.) In particular, if $\Gamma$ was a small cancellation presentation (see Section 6.1.2) to begin with, then after applying several transformations (T4′) it will, most likely, no longer be. As a result, Eve's chances to augment the public presentation $\hat{\Gamma}$ to a small cancellation presentation (see Section 6.1.5) are getting slimmer.

One more trick that Alice can use for better diffusion of her presentation is making a free product of her group with the trivial group given by a nonstandard presentation. That means, she can add new generators $z_1, \ldots, z_q$ and new relators $s_1(z_1, \ldots, z_q), \ldots, s_t(z_1, \ldots, z_q)$, such that the presentation $\langle z_1, \ldots, z_q \mid s_1, \ldots, s_t \rangle$ defines the trivial group. After that, she has to apply several (T3)s and (T4′)s to mix the new generators with the old ones. We note that there are many nontrivial presentations of the trivial group to choose from; for example, in [110], there are given several infinite series of such presentations in the special case where $t = q$ (so-called *balanced presentations*). Without this restriction, there are even more choices; in particular, Alice can just add arbitrary relators to a balanced presentation of the trivial group, thus adding to the confusion of the adversary.

After Alice has mixed $\Gamma$ by using these tricks, we suggest that she should aim for breaking down some of the defining relators into "small pieces". More formally, she can replace a given presentation by an isomorphic presentation where most defining relators have length at most 4. (Intuitively, diffusion of elements should be easier to achieve in a group with shorter defining relators.) This is easily achieved by applying transformations (T1) (see below) which can be "seasoned" by a few elementary automorphisms (type (T3)) of the form $x_i \to x_i x_j^{\pm 1}$ or $x_i \to x_j^{\pm 1} x_i$, for better diffusion.

The procedure of breaking down defining relators is quite simple. Let $\Gamma$ be a presentation $\langle x_1, \ldots, x_k; r_1, \ldots, r_m \rangle$. We are going to obtain a different, isomorphic, presentation by using Tietze transformations of types (T1). Specifically, let, say, $r_1 = x_i x_j u$, $1 \leq i, j \leq k$. We introduce a new generator $x_{k+1}$ and a new relator $r_{m+1} = x_{k+1}^{-1} x_i x_j$. The presentation $\langle x_1, \ldots, x_k, x_{k+1}; r_1, \ldots, r_m, r_{m+1} \rangle$ is obviously isomorphic to $\Gamma$. Now if we replace $r_1$ with $r_1' = x_{k+1} u$, then the presentation $\langle x_1, \ldots, x_k, x_{k+1}; r_1', \ldots, r_m, r_{m+1} \rangle$ will again be isomorphic to $\Gamma$, but now the length of one of the defining relators ($r_1$) has decreased by 1. Continuing in this manner, Alice can eventually obtain a presentation where many relators have length at most 3, at the expense of introducing more generators. In fact, relators of length 4 are also good for the purpose of diffusing a given word, so we are not going to "cut" the relators into too small pieces (i.e., we do not want pieces of length 1 or 2), but rather settle with relators of length 3 or 4. Most of the longer relators can be discarded from the presentation $\Gamma'$ to obtain the abridged presentation $\hat{\Gamma}$.

We conclude this section with a simple example, just to illustrate how Tietze transformations can be used to cut relators into pieces. In this example, we start with a presentation having two relators of length 5 in 3 generators, and end up with a presentation having four relators of length 3 or 4 in five generators. The symbol $\cong$ below means "is isomorphic to".

**Example 6.1.1.**

$$\langle x_1, x_2, x_3 \mid x_1^2 x_2^3, x_1 x_2^2 x_1^{-1} x_3 \rangle$$
$$\cong \langle x_1, x_2, x_3, x_4 \mid x_4 = x_1^2, x_4 x_2^3, x_1 x_2^2 x_1^{-1} x_3 \rangle$$
$$\cong \langle x_1, x_2, x_3, x_4, x_5 \mid x_5 = x_1 x_2^2, x_4 = x_1^2, \ x_4 x_2^3, \ x_5 x_1^{-1} x_3 \rangle$$
$$\cong \langle x_1, x_2, x_3, x_4, x_5 \mid x_5 = x_2^2, x_4 = x_1^2, \ x_4 x_2^3, \ x_1 x_5 x_1^{-1} x_3 \rangle.$$

The last isomorphism illustrates applying a transformation of type (T3), namely, the automorphism $x_5 \to x_1 x_5$, $x_i \to x_i$, $i \neq 5$.

## 6.1.4  Generating random elements in finitely presented groups

In this section, we explain how to implement the crucial step (3) of the protocol given in Section 6.1.1.

When Bob wants to transmit an element equal to 1 in $\hat{\Gamma}$, he should construct a word, looking "as random as possible" (for semantic security), in the relators $\hat{r}_1, \ldots, \hat{r}_l$ and their conjugates. When he wants to transmit an element not equal to 1 in $\hat{\Gamma}$, he just selects a random word of sufficiently big length; it turns out that, with overwhelming probability, such an element is not equal to 1 in $\Gamma'$ (we explain it in the end of this section).

We start with a description of Bob's possible diffusion strategy for producing elements equal to 1 in $\hat{\Gamma}$. (It is rather straightforward to produce a random word of a given length in generators $x_1, \ldots, x_k$, so we are not going to discuss it here.) When Bob transmits a word $w$ equal to 1 in $\hat{\Gamma}$, he wants to diffuse it so that large pieces of defining relators would not be visible in $w$. In some specific groups (e.g., in braid groups) diffusion is provided by a "normal form" (see Section 2.6), which is a collection of symbols that uniquely corresponds to a given element of the group. The existence of such normal forms is usually due to some special algebraic or geometric properties of a given group.

However, since Bob does not know any meaningful properties of the group defined by the presentation $\hat{\Gamma}$ which is given to him, he cannot employ normal forms in the usual sense. The only useful property that the presentation $\hat{\Gamma}$ has is that most of its defining relators have length 3 or 4, see Section 6.1.2. We are going to take advantage of this property as follows. We suggest the following procedure, which is probably best described by the word "shuffling".

1. Make a product of the form $u = s_1 \cdots s_p$, where each $s_i$ is randomly chosen among defining relators $\hat{r}_1, \hat{r}_2, \ldots,$ of length 3 or 4, their inverses, and their conjugates by one- or two-letter words in $x_1', x_2', \ldots$. The number $p$ of factors should be sufficiently big, at least 10 times the number of defining relators in $\hat{\Gamma}$.

2. Insert approximately $\frac{2p}{k}$ expressions of the form $x_j'(x_j')^{-1}$ or $(x_j')^{-1}x_j'$ in random places of the word $u$ (here $k$ is the number of generators $x_i'$ of $\hat{\Gamma}$), for random values of $j$.

3. Going left to right in the word $u$, look for two-letter subwords that are parts of defining relators $\hat{r}_i$ of length 3 or 4. When you spot such a subword, replace it by the inverse of the augmenting part of the same defining relator and continue. For example, suppose there is a relator $\hat{r}_i = x_1'x_2'x_3'x_4'$, and suppose you spot the subword $x_1'x_2'$ in $u$. Then replace it by $(x_4')^{-1}(x_3')^{-1}$ (obviously, $x_1'x_2' = (x_4')^{-1}(x_3')^{-1}$ in your group). If you spot the subword $x_2'x_3'$ in $w$, replace it by $(x_1')^{-1}(x_4')^{-1}$. If there is more than one choice for replacement, choose randomly among them.

4. Cancel remaining subwords (if any) of the form $x_j'(x_j')^{-1}$ or $(x_j')^{-1}x_j'$.

Steps (2)–(4) should be repeated approximately $p$ times for good mixing.

Finally, after Bob has obtained a word $u$ this way, he sets $w = [x_i', u]$ and applies steps (2)–(4) to $w$ approximately $\frac{|w|}{2}$ times, where $|w|$ is the length of $w$. This final step is needed to make this $w$ (which is equal to 1 in $\hat{\Gamma}$, and therefore also in $\Gamma'$) indistinguishable from $w \neq 1$, which is constructed in the same form $w = [x_i', u]$, see below. Here $x_i'$ is the same as in the relator $x_i' = \prod_{j=1}^{M} [x_i', w_j]$ published by Alice, see Section 6.1.2. Having $w \neq 1$ in this form is needed, in turn, to foil "quotient attacks", see the end of Section 6.1.6.

When Bob wants to transmit an element not equal to 1 in $\Gamma'$, he should first choose a random word $u$ from the commutator subgroup of the free group generated by $x_1', x_2', \ldots$. To select a random word from the commutator subgroup is easy; Bob can select an arbitrary random word $v$ first, and then adjust the exponents on the generators in $v$ so that the exponent sum on every generator in $v$ is 0. The length of $u$ should be in the same range as the lengths of the words $u$ equal to 1 in $\hat{\Gamma}$ constructed by Bob before. Then Bob lets $w = [x_i', u]$, where $x_i'$ is the same as in the relator $x_i' = \prod_{j=1}^{M} [x_i', w_j]$ published by Alice, see Section 6.1.2. Finally, to hide $u$, he applies "shuffling" to $w$ (steps (2)–(4) above) approximately $\frac{|w|}{2}$ times, where $|w|$ is the length of $w$.

Now we explain why a random word of sufficiently big length is not equal to 1 in $\Gamma$ with overwhelming probability, provided $\Gamma$ is a presentation described in the end of Section 6.1.2.

Like any other group, the group $G$ given by the presentation $\Gamma$ is a factor group $G = F/R$ of the ambient free group $F$ generated by $x_1, x_2, \ldots$. Therefore, to estimate the probability that a random word in $x_1, x_2, \ldots$ would not belong to $R$ (and therefore, would not be equal to 1 in $G$), one should estimate the *asymptotic density* (see e.g., [75]) of the complement to $R$ in the free group $F$. It makes notation simpler if one deals instead with the asymptotic density of $R$ itself, which is

$$\rho_F(R) = \limsup_{n \to \infty} \frac{\#\{u \in R : |u| \leq n\}}{\#\{u \in F : |u| \leq n\}}.$$

Here $|u|$ denotes the usual lexicographic length of $u$ as a word in $x_1, x_2, \ldots$. Thus, the asymptotic density depends, in general, on a free generating set of $F$, but we will not go into these details here because all facts that we are going to need are independent of the choice of basis. One principal fact that we can use here is due to Woess [141]: if the group $G = F/R$ is infinite, then $\rho_F(R) = 0$. Since the group $G$ given by the presentation $\Gamma$ is infinite (see our Section 6.1.2), this already tells us that the probability for a random word of length $n$ in $x_1, x_2, \ldots$ not to be equal to 1 in $G$ is approaching 1 when $n \to \infty$. However, if we want words transmitted by Bob to be of reasonable length (on the order of 100–200, say), we have to address the question of *how fast* the ratio in the definition of the asymptotic density converges to 0 if $R$ is the normal closure of the relators described in the end of Section 6.1.2. It turns out that for *non-amenable* groups the convergence is exponentially fast; this is also due to Woess [141]. We are not going to explain here what amenable groups are; it is sufficient for us to know that small cancellation groups are not amenable (because they have free subgroups, see e.g., [95]). Thus, small cancellation groups are just fine for our purposes here: the probability for a random word of length $n$ in $x_1, x_2, \ldots$ not to be equal to 1 in $G$ is approaching 1 exponentially fast when $n \to \infty$.

Finally, we touch upon semantic security (see [50]) of the words transmitted by Bob. We do not give any rigorous probabilistic estimates since this would require at least defining a probability measure on an infinite group, which is a very nontrivial problem by itself (cf. [16]). Instead, we offer here an informal argument which we hope to be convincing. A nice thing about Bob's encryption procedure is that when he selects a word $u \neq 1$, he simply selects a random word. Thus, $u \neq 1$ is indistinguishable from a random word just because it is random! Then, the element $w = [x'_i, u]$, which is transmitted by Bob, looks like it is no longer random because it is of a special form. However:

1. What is actually transmitted by Bob is a *word in the alphabet* $x'_1, x'_2, \ldots$ representing the element $w = [x'_i, u]$ of the group defined by $\hat{\Gamma}$. This word is *not* of the form $[x'_i, u]$ because Bob has applied a "shuffling" to $w$.

2. Given the specifics of our protocol, what really matters is that transmitted words equal to 1 in $\hat{\Gamma}$ are indistinguishable from transmitted words not equal to 1. This is why we require Bob's elements representing 1 in $\hat{\Gamma}$ to be of the form $[x'_i, u]$ as well.

Thus, the question about semantic security of Bob's transmissions boils down to the following question of independent interest: is a word $u$ representing 1 in $\hat{\Gamma}$ indistinguishable from a random word (of the same length)? As we have admitted above, we do not have a rigorous proof that it is, but computer experiments show that when most of the relators in $\hat{\Gamma}$ have length at most 4, then the words $u$ representing 1 in $\hat{\Gamma}$, obtained as described earlier in this section, pass at least the equal frequency test for 1-, 2-, and 3-letter subwords, thus making it appear likely that the answer to the question above is affirmative for such $\hat{\Gamma}$.

## 6.1.5  Isomorphism attack

In this section, we discuss a (theoretically) possible "brute force" attack on the protocol from Section 6.1.1.

Knowing the pool of group presentations from which Alice selects her private presentation $\Gamma$, Eve can try to augment the public presentation $\hat{\Gamma}$ to a presentation that would be isomorphic to one from the pool. Theoretically, this is possible because the pool is recursive and because the set of finite presentations isomorphic to a given one is recursive, too. However, this procedure requires enormous resources. Let us take a closer look at it.

Eve can add to $\hat{\Gamma}$ one element at a time and check whether the resulting presentation, call it $\hat{\Gamma}_+$, is isomorphic to one of the presentations from Alice's pool. The latter is done the following way. Suppose Eve wants to check whether $\hat{\Gamma}_+$ is isomorphic to some $\Gamma_i$. She goes over mappings from $\Gamma_i$ to $\hat{\Gamma}_+$, one at a time, defined on the generators of $\Gamma_i$. At the same time, she also goes over mappings from $\hat{\Gamma}_+$ to $\Gamma_i$ defined on the generators of $\hat{\Gamma}_+$. She composes various pairs of these mappings and checks: (1) whether she gets the identical mapping on $\Gamma_i$, and (2) whether both mappings in such a pair are homomorphisms, i.e., whether they send relators of either presentation to elements equal to 1 in the other presentation. Having the word problem in $\Gamma_i$ solvable makes the former checking more efficient, but it is, in fact, not necessary because what matters here is the "yes" part of the word problem, which is always recursive.

Now let us focus on the part of this procedure where Eve works with a particular presentation $\Gamma_i$ from Alice's pool. Suppose $\Gamma_i$ is not isomorphic to $\hat{\Gamma}_+$. Since the "no" part of the isomorphism problem between $\hat{\Gamma}_+$ and $\Gamma_i$ is not recursive, Eve would have to try out various pairs of mappings between $\hat{\Gamma}_+$ and $\Gamma_i$ (see above) indefinitely. Therefore, she will have to allocate (indefinitely) some memory resources to checking this particular $\Gamma_i$. Since the number of $\Gamma_i$ grows exponentially with the size of the presentation (which is the total length of relators), Eve would require essentially unlimited storage space and, in fact, she will reach physical limits (e.g., the number of electrons in the universe) on the storage space very quickly because, say, the number of presentations on six generators with the total length of relators bounded by 100 is already more than $10^{100}$.

We make a disclaimer saying that there may be smarter ways to find a small cancellation presentation isomorphic to $\hat{\Gamma}_+$, but we hope we have convinced the

reader that (at least, in the worst case) this search would require essentially unlimited computational resources.

## 6.1.6   Quotient attack

In this section, we discuss an attack which is, in general, more efficient (especially
in real life) than the "brute force" attack described in Section 6.1.5. We use here
some group-theoretic terminology not supported by formal definitions when we
feel it should not affect the reader's understanding of the material. Some of the
basic terminology has to be recalled though.

A group $G$ is called *abelian* (or commutative) if $[a, b] = 1$ for any $a, b \in G$,
where $[a, b]$ is the notation for $a^{-1}b^{-1}ab$. Thus, $[a, b] = 1$ is equivalent to $ab = ba$.
This can be generalized in different ways. A group $G$ is called *metabelian* (see
also our Section 5.5) if $[[x, y], [z, t]] = 1$ for any $x, y, z, t \in G$. A group $G$ is called
*nilpotent of class* $c \geq 1$ if $[y_1, y_2, \ldots, y_{c+1}] = 1$ for any $y_1, y_2, \ldots, y_{c+1} \in G$, where
$[y_1, y_2, y_3] = [[y_1, y_2], y_3]$, etc.

We note that in the definition of an abelian group, it is sufficient to require
that $[x_i, x_j] = 1$ for all *generators* $x_i, x_j$ of the group $G$. Thus, *any finitely generated abelian group is finitely presented.* The same is true for all finitely generated
nilpotent groups of any class $c \geq 1$, but not for all metabelian groups. In particular, it is known that finitely generated *free metabelian groups* are infinitely
presented [6]. A free metabelian group is the factor group of a free group by the
second commutator subgroup, see our Section 5.5.

Now we get to quotient attacks. One way for Eve to try to positively identify
those places in Bob's binary sequence where he intended to transmit a 0 is to use
a *quotient test* (see e.g., [75] for a general background). That means the following:
Eve tries to add finitely or infinitely many relators to the given presentation $\hat{\Gamma}$
to obtain a presentation defining a group $H$ with solvable word problem (more
accurately, a group $H$ that Eve can *recognize* as having solvable word problem).

It makes sense for Eve to only try recognizable quotients, such as, for example,
abelian or, more generally, nilpotent ones. This amounts to adding specific relators
to $\hat{\Gamma}$; for example, for an abelian quotient, Eve can add relators $[x_i', x_j']$ for all pairs
of generators $x_i', x_j'$ in $\hat{\Gamma}$. For nilpotent quotients, Eve will have to add commutators
of higher weight in the generators. For a metabelian quotient, Eve will have to add
infinitely many relators (because, as we have already mentioned, free metabelian
groups are infinitely presented), but this is not a problem since she does not have
to "actually add" those relators; she can just consider $\hat{\Gamma}$ as a presentation *in
the variety of metabelian groups* and apply the relevant algorithm for solving the
word problem which is universal for all groups finitely presented in the variety of
metabelian groups, see [81] for more details.

Note that this trick will *not* work with hyperbolic quotients, say. This is
because there is no way, in general, to add specific relators to $\hat{\Gamma}$ to make sure that
the extended presentation defines a hyperbolic group. This deprives Eve from using

a (rather powerful, cf. [75]) hyperbolic quotient attack.

Classes of groups with solvable word problem are summarized in the survey [81]. It appears that a quotient attack can essentially employ either a nilpotent or a metabelian quotient of $\hat{\Gamma}$. This is why, to foil such attacks, Alice adds a relator $x_i' = \prod_{j=1}^{M}[x_i', w_j]$ to $\hat{\Gamma}$ (see our Section 6.1.2). This is also the reason why Bob should choose a word of the form $[x_i', u]$ when he wants to transmit an element not equal to 1 in $\Gamma'$ (see Section 6.1.4). Indeed, a metabelian quotient attack on an element of the form $[x_i', u]$ will not work because this element belongs to the second commutator subgroup of the group defined by $\hat{\Gamma}$ since in this group, $x_i' = \prod_{j=1}^{M}[x_i', w_j]$, so $x_i'$ belongs to the commutator subgroup of the given group. Furthermore, an element of the form $[x_i', u]$ belongs to every term of the lower central series of the given group since in this group, $[x_i', u] = [\prod_{j=1}^{M}[x_i', w_j], u] = [\prod_{j=1}^{M}[\prod_{j=1}^{M}[x_i', w_j], w_j], u]$, etc. This foils nilpotent quotient attacks, too.

## 6.2 Public key encryption and encryption emulation attacks

In this section we continue to explore how non-recursiveness of a decision problem (as opposed to computational hardness of a search problem) can be used in public key cryptography. We follow the exposition in [113].

Here our focus is on the "encryption emulation" attack on the sender's (Bob's) transmissions. We show that Bob's encryption can be made reasonably secure against the encryption emulation attack (see our Section 6.1.1) by computationally unbounded adversary, with one reservation: a legitimate receiver decrypts correctly with probability that can be made arbitrarily close to 1, but not equal to 1.

First we recall (from Section 6.1.1) that the encryption emulation attack would indeed work fine (at least, for a computationally unbounded adversary) if the correctness of the scheme was perfect. However, if there is a gap, no matter how small (it can be easily made on the order of $10^{-200}$), between 1 and the probability of correct decryption by a legitimate receiver, then this gap can be very substantially "amplified" for the adversary, thus making the probability of correct illegitimate decryption anything but overwhelming.

We emphasize at this point that in the protocol described in this section, we do not claim security against the "encryption emulation" (or any other) attack by a computationally unbounded adversary on the receiver's public key, but we only claim security against the "encryption emulation" attack on the sender's transmission. It seems that the problem of security of the sender's encryption algorithm is of independent interest. Of course, in applications to, say, Internet shopping or banking, both the sender's and the receiver's algorithms are assumed to be known to the adversary ("Kerckhoffs' assumptions", see e.g., [100]), and the receiver's decryption algorithms (or algorithms for obtaining public keys) are usually more vulnerable to attacks. However, in some other applications, say, to electronic signatures (not to mention non-commercial, e.g., military applications), decryption algorithms or algorithms for generating public keys (by the receiver) need not be public, whereas encryption algorithms (of the sender) always are.

Thus, we present here a protocol which is secure against the "encryption emulation" attack on the sender's transmission by a computationally unbounded adversary who has complete information on the algorithm(s) and hardware that the sender uses for encryption. More precisely, in our protocol the sender transmits his private bit sequence by encrypting one bit at a time, and the receiver decrypts each bit correctly with probability that can be made arbitrarily close to 1, but not equal to 1. At the same time, the (computationally unbounded) adversary decrypts each bit (by emulating the sender's encryption algorithm) correctly with probability at most $\frac{3}{4}$.

There are essentially no requirements on the sender's computational abilities; in fact, encryption can be done by hand, which can be a big advantage in some situations; for example, a field operative can receive a public key from a command center and transmit encrypted information over the phone, without even using a computer.

**Encryption.**    Now we describe an encryption protocol with the following features:

(F1)  Bob encrypts his secret bit by a word in a public alphabet $X$.

(F2)  Alice (the receiver) decrypts Bob's transmission correctly with probability that can be made arbitrarily close to 1, but not equal to 1.

(F3)  The adversary, Eve, is assumed to have no bound on the speed of computation or on the storage space.

(F4)  Eve is assumed to have complete information on the algorithm(s) and hardware that Bob uses for encryption. However, Eve cannot predict outputs of Bob's random numbers generator (the latter could be just coin tossing, say).

(F5)  Eve does not have information on Alice's algorithm for obtaining public keys.

(F6)  Eve cannot decrypt Bob's secret bit correctly with probability $> \frac{3}{4}$ by emulating Bob's encryption algorithm.

Once again: here we only claim security against the "encryption emulation" attack (by a computationally unbounded adversary) on the sender's transmissions. This does not mean that the receiver's private keys in our protocol are insecure against real-life (i.e., computationally bounded) adversaries, but we just prefer to focus here on what is secure against a computationally unbounded adversary since this paradigm shift looks important to us (at least, from a theoretical point of view).

We also have to say up front that the encryption protocol that is presented in this section is probably not very suitable for commercial applications (such as

Internet shopping or banking) for yet another reason: because of a large amount of work required from Alice to receive just one bit from Bob. Bob, on the other hand, may not even need a computer for encryption.

Now we are getting to the protocol description. In one round of this protocol, Bob transmits a single bit, i.e., Alice generates a new public key for each bit transmission.

(P0) Alice publishes two group presentations by generators and defining relators:

$$\Gamma_1 = \langle x_1, x_2, \ldots, x_n \mid r_1, r_2, \ldots, r_k \rangle,$$

$$\Gamma_2 = \langle x_1, x_2, \ldots, x_n \mid s_1, s_2, \ldots, s_m \rangle.$$

One of them defines the trivial group, whereas the other one defines an infinite group, but only Alice knows which one is which. In the group that is infinite, Alice should be able to efficiently solve the word problem, i.e., given a word $w = w(x_1, x_2, \ldots, x_n)$, she should be able to determine whether or not $w = 1$ in that group. (There is a large and easily accessible pool of such groups, see our Section 6.1.2 for discussion.)

Bob is instructed to transmit his private bit to Alice as follows:

(P1) In place of "1", Bob transmits a pair of words $(w_1, w_2)$ in the alphabet $X = \{x_1, x_2, \ldots, x_n, x_1^{-1}, \ldots, x_n^{-1}\}$, where $w_1$ is selected randomly, while $w_2$ is selected to be equal to 1 in the group $G_2$ defined by $\Gamma_2$.

(P2) In place of "0", Bob transmits a pair of words $(w_1, w_2)$, where $w_2$ is selected randomly, while $w_1$ is selected to be equal to 1 in the group $G_1$ defined by $\Gamma_1$.

Now we have to specify the algorithms that Bob should use to select his words.

**Algorithm "0"** (for selecting a word $v = v(x_1, \ldots, x_n)$ not equal to 1 in a $\Gamma_i$) is quite simple: Bob just selects a random word by building it letter-by-letter, selecting each letter uniformly from the set $X = \{x_1, \ldots, x_n, x_1^{-1}, \ldots, x_n^{-1}\}$. The length of such a word should be a random integer from an interval that Bob selects up front, based on his computational abilities. In the end, Bob should cancel out all subwords of the form $x_i x_i^{-1}$ or $x_i^{-1} x_i$.

**Algorithm "1"** (for selecting a word $u = u(x_1, \ldots, x_n)$ equal to 1 in a $\Gamma_i$) is slightly more complex. It amounts to applying a random sequence of operations of the following two kinds, starting with the empty word:

1. Inserting into a random place in the current word a pair $hh^{-1}$ for a random word $h$.

2. Inserting into a random place in the current word a random conjugate $g^{-1} r_i g$ of a random defining relator $r_i$.

In the end, Bob should cancel out all subwords of the form $x_i x_i^{-1}$ or $x_i^{-1} x_i$. The length of the resulting word should be in the same range as the length of the output of Algorithm "0". We do not go into more details here because all claims in this section remain valid no matter what algorithm for producing words equal to 1 is chosen, as long as it returns a word whose length is in the same range as that of the output of Algorithm "0".

Now let us explain why the legitimate receiver (Alice) decrypts correctly with overwhelming probability. Suppose, without loss of generality, that the group $G_1$ is trivial, and $G_2$ is infinite. Then, if Alice receives a pair of words $(w_1, w_2)$ such that $w_1 = 1$ in $G_1$ and $w_2 \neq 1$ in $G_2$, she concludes that Bob intended to transmit a "0". This conclusion is correct with probability 1. If Alice receives $(w_1, w_2)$ such that $w_1 = 1$ in $G_1$ and $w_2 = 1$ in $G_2$, she concludes that Bob intended to transmit a "1". This conclusion is correct with probability which is close to 1, but not equal to 1 because it may happen, with probability $\varepsilon > 0$, that the random word $w_2$ selected by Bob is equal to 1 in $G_2$. The point here is that, if $G_2$ is infinite, this $\varepsilon$ is negligible and, moreover, for "most" groups $G_2$ this $\varepsilon$ tends to 0 exponentially fast as the length of $w_2$ increases. For more precise statements, see our Section 6.1.4; here we just say that it is easy for Alice to make sure that $G_2$ is one of those groups.

Now we are going to discuss Eve's attack on Bob's transmission. Under our assumptions (F3), (F4) Eve can identify the word(s) in the transmitted pair which is/are equal to 1 in the corresponding group(s), as well as the word, if any, which is not equal to 1. Indeed, for any particular transmitted word $w$ she can use the "encryption emulation" attack, as described in our Introduction: she emulates algorithms '0' and "1" over and over again, each time with fresh randomness, until the word $w$ is obtained. Thus, Eve is building up two lists, corresponding to two algorithms above. As we have already pointed out in a similar situation in Section 6.1.1, the list that corresponds to the Algorithm "0" is useless to Eve because it is eventually going to contain *all* words in the alphabet $X = \{x_1, \ldots, x_n, x_1^{-1}, \ldots, x_n^{-1}\}$, with overwhelming probability. Therefore, Eve may just as well forget about this list and concentrate on the other one, that corresponds to the Algorithm "1". Now the situation boils down to the following: if the word $w$ appears on the list, then it is equal to 1 in the corresponding group $G_i$. If not, then not.

It may seem that Eve should encounter a problem detecting $w \neq 1$: how can she conclude that $w$ does *not* appear on the list if the list is infinite (more precisely, of a priori unbounded length) ? This is where our condition (F4) plays a role: if Eve has complete information on the algorithm(s) and hardware that Bob uses for encryption, then she does know a real-life bound on the size of the list.

Thus, Eve can identify the word(s) in the transmitted pair which is/are equal to 1 in the corresponding group(s), as well as the word, if any, which is not equal to 1. There are the following possibilities now:

1. $w_1 = 1$ in $G_1$, $w_2 = 1$ in $G_2$;

2. $w_1 = 1$ in $G_1$, $w_2 \neq 1$ in $G_2$;

3. $w_1 \neq 1$ in $G_1$, $w_2 = 1$ in $G_2$.

It is easy to see that one of the possibilities (2) or (3) cannot actually occur, depending on which group $G_i$ is trivial. Then, the possibility (1) occurs with probability $\frac{1}{2}$ (either when Bob wants to transmit "1" and $G_1$ is trivial, or when Bob wants to transmit "0" and $G_2$ is trivial). If this possibility occurs, Eve cannot decrypt Bob's bit correctly with probability $> \frac{1}{2}$ because she does not know which group $G_i$ is trivial. As we have pointed out earlier in this section, a computationally unbounded Eve could find out which $G_i$ is trivial, but we specifically consider attacks on the sender's encryption here (cf., our condition (F5) above). We just note, in passing, that for a real-life (i.e., computationally bounded) adversary to find out which presentation $\Gamma_i$ defines the trivial group is by no means easy and deserves to be a subject of separate investigation. There are many different ways to efficiently construct very complex presentations of the trivial group, some of them involving a lot of random choices. See e.g., [110] for a survey on the subject.

In any case, our claim (F6) was that Eve cannot decrypt Bob's bit correctly with probability $> \frac{3}{4}$ by emulating Bob's encryption algorithm, which is obviously true in this scheme since the probability for Eve to decrypt correctly is, in fact, precisely $\frac{1}{2} \cdot \frac{1}{2} + \frac{1}{2} \cdot 1 = \frac{3}{4}$. (Note that Eve decrypts correctly with probability 1 if either of the possibilities (2) or (3) above occurs.)

Someone may say that $\frac{3}{4}$ is a rather high probability of illegitimate decryption, even though this is just for one bit. Recall however that we are dealing with a computationally unbounded adversary, while Bob can essentially do his encryption by hand! All he needs is a generator of uniformly distributed random integers in the interval between 1 and $2n$ (the latter is the cardinality of the alphabet $X$). Besides, note that with the probability of correctly decrypting one bit equal to $\frac{3}{4}$, the probability of correctly decrypting, say, a credit card number of 16 decimal digits would be on the order of $10^{-7}$, which is comparable to the chance of winning the jackpot in a lottery. Of course, there are many tricks that can make this probability much smaller, but we think we better stop here because, as we have pointed out before, our focus here is on the new paradigm itself.

# Part III

# Generic Complexity and Cryptanalysis

In this part of the book, we discuss two measures of complexity of an algorithm, average-case and generic-case, and argue that it is the generic-case complexity (we often call it just generic complexity) that should be considered in the context of cryptanalysis of various cryptographic schemes.

Since in our book we mostly deal with cryptographic schemes based on non-commutative infinite groups, we give a survey of what is known about generic-case behavior of various algorithms typically studied in combinatorial group theory and relevant to cryptography. The very idea of "genericity" in group theory was introduced by Gromov and Ol'shanskii and is now a subject of very active research. Genericity exhibits itself on many different levels in algebraic and algorithmic properties of "random" algebraic objects and in the generic-case behavior of their natural geometric invariants. A generic approach often leads to the discovery of objects with genuinely new and interesting algebraic properties. For example, genericity provides a totally new source of *group-theoretic rigidity*, quite different from the standard source provided by lattices in semisimple Lie groups. Generic-case analysis of group-theoretic algorithms gives a much more detailed and stratified "anatomical" picture of an algorithm than worst-case analysis provides. It also gives a better measure of the overall "practicality" of an algorithm and often leads to genuine *average-case* results.

It is probably safe to say that by now it is pretty clear that the worst-case complexity is not necessarily a good measure of "practicality" of an algorithm. The paradigm example is Dantzig's Simplex Algorithm for linear programming. Very clever examples of Klee and Minty [82] show that the Simplex Algorithm can be made to take exponential time. However, the simplex method runs thousands times a day in many real-life applications and it always works very fast, namely in linear time. Although there are provably polynomial time algorithms for linear programming, these have not replaced the Simplex Algorithm in practice. It turns out that this type of phenomenon, which we refer to as having low "generic-case complexity" is very pronounced in group theory. This notion was introduced and formalized by Kapovich, Myasnikov, Schupp and Shpilrain in [75]. In particular, they showed that for most groups usually studied in combinatorial or geometric group theory, the generic-case complexity of the word problem is linear time, often in a very strong sense. Consequently, the actual average-case complexity is also often linear time.

We note that unlike the average-case complexity, generic-case complexity completely disregards potentially bad behavior of an algorithm on an asymptotically negligible set of inputs. Thus generic-case complexity of an algorithm $\mathcal{A}$ may be low (e.g., linear time) even if the relevant algorithmic problem is undecidable (and $\mathcal{A}$ has infinite running time on some inputs), while the average-case complexity would necessarily be infinite in this case. As was observed in [75], this is exactly what happens with the classical decision problems in group theory, such as the word, conjugacy and subgroup membership problems. Moreover, in [76] the same authors were also able to apply their generic-case methods to obtain genuine average-case complexity conclusions. The basic idea there is that by running in parallel a total algorithm with subexponential worst-case complexity and a partial algorithm with low *strong generic-case* complexity, one obtains an algorithm with low average-case complexity. Typical results of [76] imply, for example, that for all braid groups, all groups of knots with hyperbolic complements and all Artin groups of extra large type, the average-case complexity of the word problem is linear time.

Apart from shedding light on "practicality" of various algorithms, generic-case analysis has some additional benefits. First, it provides a natural stratification of the set of inputs of a problem into the "easy" and "not necessarily easy" parts. Concentration on the latter part often forces one to redefine the original problem and substantially change its nature. By iterating this process one can obtain a detailed analysis of the "anatomy" of an algorithmic problem that is in many ways more informative than the worst-case analysis. Isolating the "hard" part of an algorithmic problem may be important for possible practical applications, in particular to cryptanalysis of various cryptographic schemes. This is what we explore in more detail in part IV of this book.

Moreover, as our experience shows, generic-case considerations can lead to the discovery of new algorithms with better average-case and worst-case behavior.

Finally, we should emphasize that one must not confuse the fact that generic results are often easy to state with simplicity of their proof. Proofs usually require interaction of some aspects of probability theory with deep results about algebraic questions under consideration.

# Chapter 7

# Distributional Problems and the Average-Case Complexity

## 7.1 Distributional computational problems

One of the aims of the section is to set up a framework that would allow one to analyze behavior of algorithms at large, to study expected or average properties of algorithms, or its behavior on "most" inputs. The key point here is to equip algorithmic problems with probability distributions on their sets of instances.

### 7.1.1 Distributions and computational problems

To study behavior of algorithms on average or on "most" inputs one needs to have a distribution on the set of inputs. In this section we discuss typical distributions that occur on discrete sets of inputs.

**Definition 7.1.1.** A *distributional computational problem* is a pair $(\mathcal{D}, \mu)$ where $\mathcal{D} = (L, I)$ is a computational problem and $\mu$ is a probability measure on $I$.

The choice of $\mu$ is very important, it should be natural in the context of the problem. Quite often, the set $I$ is infinite and discrete (enumerable), so, unlike in the finite set case, it is impossible to use uniform distributions on $I$. Discreteness of the set $I$ usually implies that the probability distributions $\mu$ are *atomic*, i.e., $\mu(x)$ is defined for every singleton $\{x\}$ and for a subset $S \subseteq I$,

$$\mu(S) = \sum_{x \in S} \mu(x). \tag{7.1}$$

It follows that an atomic measure $\mu$ is completely determined by its values on singletons $\{x\}, x \in I$, so to define $\mu$ it suffices to define a function $p : I \to \mathbb{R}$ (with $\mu(\{x\}) = p(x)$), which is called a *probability mass function* or a *density function* on $I$, such that $p(x) \geq 0$ for all $x \in I$, and $\sum_{x \in I} p(x) = 1$.

There are two general basic ideas on how to introduce a natural distribution on $I$.

1. *Distributions via generating procedures.* If elements of $I$ can be naturally generated by some "random generator" $R$, then probability $\mu(x)$ can be defined as the probability of the generator $R$ to produce $x$. We refer to [105] for a detailed discussion.

2. *Distributions via stratifications.* Let $s : I \to \mathbb{N}$ be a size function on $I$. An atomic distribution $\mu$ on $I$ *respects the size* function $s$ if:

$$\forall x, y \in I \quad (s(x) = s(y) \longrightarrow \mu(x) = \mu(y)).$$

Such a measure $\mu$ on $I$ is termed *size invariant* or *homogeneous* or *uniform*. More about homogeneous measures can be found in [5]. Here we give only some facts that are required in the sequel.

Observe, that a homogeneous distribution $\mu$ induces a uniform distribution on each sphere $I_k$, so if the induced distribution is nonzero, then the sphere is finite. The following result describes size-invariant distributions on $I$.

**Lemma 7.1.2.** *Let $\mu$ be an atomic measure on $I$ and $s$ a complexity function on $I$ with finite spheres $I_k$. Then:*

(1) *If $\mu$ is $s$-invariant, then the function $d_\mu : \mathbb{N} \longrightarrow \mathbb{R}$ defined by*

$$d_\mu : k \longrightarrow \mu(I_k)$$

*is an atomic probability measure on $\mathbb{N}$;*

(2) *if $d : \mathbb{N} \longrightarrow \mathbb{R}$ is an atomic probability measure on $\mathbb{N}$, then the function $p_{s,d} : I \longrightarrow \mathbb{R}$ defined by*

$$p_{s,d}(x) = \frac{d(s(x))}{|I_n|}$$

*is a probability density function which gives rise to an atomic $s$-invariant measure on $I$.*

*Proof* is obvious.                                                                         □

**Remark 7.1.3.** Since $d(k) \to 0$ as $k \to \infty$, the more complex (with respect to a given size function $s$) elements of a set $R \subset I$ contribute less to $\mu_{s,d}(R)$, so there is a bias to elements of small size.

This discussion shows that the homogeneous probability measures $\mu_{s,d}$ on $I$ are completely determined by the complexity function $s : I \longrightarrow \mathcal{N}$ and a moderating distribution $d : \mathbb{N} \longrightarrow \mathbb{R}$. Here are some well-established parametric families of density functions on $\mathbb{N}$ that will occur later on in this framework:

The *exponential density*:

$$d_\lambda(k) = (1 - e^{-\lambda})e^{-\lambda k}.$$

The *Cauchy density:*

$$d_\lambda(k) = b \cdot \frac{1}{(k - \lambda)^2 + 1}.$$

The *Dirac density:*

$$d_m(k) = \begin{cases} 1 & \text{if } k = m, \\ 0 & \text{if } k \neq m. \end{cases}$$

The following density function depends on a given size function $s$ on $I$:

The *finite disc uniform density:*

$$d_m(k) = \begin{cases} \frac{|I_k|}{|B_m(I)|} & \text{if } k \leq m, \\ 0 & \text{otherwise.} \end{cases}$$

For example, the Cauchy density functions are very convenient for defining degrees of polynomial growth "on average", while the exponential density functions are suitable for defining degrees of exponential growth "on average". Exponential density functions appear naturally as distributions via random walks on graphs or as *multiplicative* or *Boltzmann* distributions (see [17] for details).

Other distributions arise from different random generators of elements in $I$. For example, Dirac densities correspond to the uniform random generators on the spheres $I_n$; finite disc densities arise from uniform random generators on the discs $B_n(I)$.

## 7.1.2  Stratified problems with ensembles of distributions

In this section we discuss stratified computational problems $\mathcal{D}$ with ensembles of distributions. In this case the set of instances $I$ of the problem $\mathcal{D}$ comes equipped with a collection of measures $\{\mu_n\}$ each of which is defined on the sphere $I_n$ (or a ball $B_n$). In this case we do not assume in advance that there exists a measure on the set $I$.

Let $\mathcal{D}$ be a stratified computational problem with a set of instances $I = I_{\mathcal{D}}$ and a size function $s = s_{\mathcal{D}} : I \to \mathbb{R}$. If for every $n$ the sphere $I_n$ comes equipped with a probability distribution $\mu_n$, then the collection of measures $\mu = \{\mu_n\}$ is called a *spherical ensemble* of distributions on $I$. Similarly, a *volume ensemble* of distributions $\mu = \{\mu_n\}$ is a collection of distributions such that $\mu_n$ is a distribution on the ball $B_n$ and such that $\mu_{n-1}$ is the measure induced from $\mu_n$ on $B_{n-1}$. Here, and in all similar situations, we extend the standard definition of the induced measure $\mu_S$ to subsets $S \subseteq B_n$ of measure zero ($\mu_n(S) = 0$) defining $\mu_S(w) = 0$ for all $w \in S$.

**Example 7.1.4.** Suppose the spheres $I_n$ are finite for every $n \in \mathbb{N}$. Then the uniform distribution $\mu_n$ on $I_n$ gives a spherical ensemble of distributions $\mu = \{\mu_n\}$ on $I$.

The following is a typical way to define ensembles of distributions on $I$. We say that a probability distribution $\mu$ on $I$ is *compatible* with the size function $s : I \to \mathbb{N}$ if $s$ is $\mu$-measurable, i.e., for every $n$ the sphere $I_n$ is a $\mu$-measurable set. Now, if $\mu$ is a probability distribution on $I$ compatible with $s$, then for every $n$ such that $\mu(I_n) \neq 0$ $\mu$ induces a measure $\mu_n$ on $I_n$. We extend this definition for all $n$ defining $\mu_n(w) = 0$ for every $w \in I_n$ in the case $\mu(I_n) = 0$.

This gives an induced ensemble of spherical distributions $\{\mu_n\}$ on $I$. For instance, the probability distributions defined in Section 7.1.1 give rise to typical spherical ensembles of distributions on $I$.

In some sense the converse also holds. If $\{\mu_n\}$ is a spherical ensemble of distributions on $I$, then one can introduce a distribution $\tilde{\mu}_d$ on $I$ which induces the ensemble $\mu$. Indeed, let $d : \mathbb{N} \longrightarrow \mathbb{R}$ be a moderating distribution on $\mathbb{N}$, so $\Sigma_n d(n) = 1$. For a subset $R \subseteq I$ such that $R_n = R \cap I_n$ is $\mu_n$-measurable for every $n$ define $\tilde{\mu}_d(R)$ as

$$\tilde{\mu}_d(R) = \Sigma_n d(n)\mu_n(R_n).$$

Clearly, the series above converges since $\Sigma_n d(n)\mu_n(R_n) \leq \Sigma_n d(n) = 1$. It is easy to see that $\tilde{\mu}_d$ is a distribution on $I$ that induces the initial ensemble $\mu$ on the spheres of $I$. Notice, that if $\mu_n$ is an atomic distribution on $I_n$, then $\mu_d$ is an atomic distribution on $I$.

The argument above shows that under some natural conditions one can harmlessly switch between distributions on $I$ and the spherical ensembles of distributions on $I$. Similar constructions hold for volume ensembles as well.

## 7.1.3   Randomized many-one reductions

A very important theoretical class of reductions is many-one reductions done by a *probabilistic Turing machine* called *randomized many-one reductions*. Intuitively a probabilistic Turing machine is a Turing machine with a random number generator. More precisely it is a machine with two transition functions $\delta_1$ and $\delta_2$. At each step of computations with probability 50% the function $\delta_1$ is used and with probability 50% the function $\delta_2$ is used. So, a probabilistic Turing machine is the same as a nondeterministic Turing machine with the only difference being in how we interpret its computations. If for NTM $M$ we are interested in knowing if there exists a sequence of choices that make $M$ accept a certain input, for the same PTM $M$ we are interested in which probability acceptance happens.

**Definition 7.1.5.** Let $D$ be a decision problem. We say that a PTM $M$ decides $D$ if it outputs that right answer with probability at least $2/3$.

**Definition 7.1.6.** We say that a PTM $M$ is *polynomial time on average* if there exists $c > 0$ such that

$$\sum_{x,z} T_M^c(x,z)\mu'(x)P(z) < \infty,$$

i.e., complexity function $T_M^c(x,z)$ is polynomial time on average taken over the distribution $\mu_1$ on inputs $x$ and the distribution $P$ on internal coin tosses $z$.

**Definition 7.1.7.** Let $D_1$ and $D_2$ be decision problems. We say that a probabilistic Turing machine $M$ *randomly reduces* $D_1$ to $D_2$ if for any $x$,

$$P(z \mid x \in D_1 \Leftrightarrow M(x, z) \in D_2) \geq 2/3$$

where probability $P$ is taken over all internal coin tosses $z$.

## 7.2 Average case complexity

In this section we briefly discuss some principle notions of the average case complexity. There are several different approaches to the average case complexity, but in some sense, they all involve computing the expected value of the running time of an algorithm with respect to some measure on the set of inputs.

More generally, one can develop a theory of average case behavior of arbitrary functions $f : I \to \mathbb{R}^+$, not only the time functions of algorithms, provided the set $I$ is a probability space equipped with a size function. From this standpoint, the average case complexity of algorithms appears as a particular application of the average case analysis to the time (or space, or any other resource) functions of algorithms.

Below we assume, if not said otherwise, that $I$ is a set (of "inputs") with a size function $s : I \to \mathbb{N}$ and a probability measure $\mu$. Throughout this section we assume also that the set $I$ is *discrete* (finite or countable) and the measure $\mu$ on $I$ is *atomic*, so all functions defined on $I$ are $\mu$-measurable. Notice, however, that all the results and definitions hold in the general case as well, if the size function $s$ and all the functions in the consideration are $\mu$-measurable.

Sometimes, instead of the measure $\mu$ we consider ensembles of spherical (or volume) distributions $\{\mu_n\}$ on spheres $I_n$ (balls $B_n$) of $I$ relative to the size function $s$. This approach seems to be more natural in applications, and gives a direct way to evaluate asymptotic behavior of functions relative to the given size.

We start with polynomial on average functions (Section 7.2.1), then discuss the general case (Section 7.2.2), and apply the average case analysis to the time complexity of algorithms (Section 7.2.3).

In Section 7.2.4 we compare the average case complexity with the worst case complexity and discuss what kind of hardness the average case complexity actually measures. In the last Section 7.2.5 we focus on deficiencies of the average case complexity.

### 7.2.1 Polynomial on average functions

There are several approaches to polynomial on average functions, some of them result in equivalent definitions and some do not. Here we follow Levin's original approach [93], which has been further developed by Y.Gurevich [58] and Impagliazzo [69].

Let $f : I \to \mathbb{R}^+$ be a nonnegative real function. We say that $f$ has a polynomial upper bound with respect to the size $s$ if there exists a real polynomial $p$ such that $f(w) \le p(s(w))$ for any $w \in I$, i.e.,

$$\forall w \in I_n \quad f(w) \le p(n).$$

To get a similar condition "on average" one may integrate the inequality above, which results in the following "naive definition" of the polynomial on average functions

**Definition 7.2.1.** A function $f : I \to \mathbb{R}^+$ is expected polynomial on spheres (with respect to an ensemble of spherical distributions $\{\mu_n\}$) if for every $n \in \mathbb{N}$,

$$\int_{I_n} f(x)\mu_n(x) \le p(n).$$

Equivalently, $f$ is expected polynomial on spheres if there exists $k \ge 1$ such that

$$\int_{I_n} f(w)\mu_n(w) = O(n^k). \tag{7.2}$$

No doubt, the functions satisfying (7.2) are "polynomial on $\mu$-average" with respect to the size function $s$, but the class of these functions is too small. Indeed, one should expect that the class of polynomial on average functions must be closed under addition, multiplication, and multiplication by a scalar. The following example shows that the class of functions satisfying the condition (7.2) is not closed under multiplication.

**Example 7.2.2.** Let $I = \{0,1\}^*$ be the set of all words in the alphabet $\{0,1\}$, $s(w) = |w|$ be the length of the word, and $\mu$ be length-preserving, so $\mu_n$ is the uniform distribution on $I_n$. Denote by $S_n$ a fixed subset of $I_n$ which has $\frac{2^n-1}{2^n}$ elements. Now we define a function $f : I \to \mathbb{N}$ by its values on each sphere $I_n$ as follows:

$$f(w) = \begin{cases} n, & \text{if } w \in S_n, \\ 2^n, & \text{if } w \notin S_n. \end{cases}$$

Then

$$\int_I f(w)\mu(w) = n(1 - 2^{-n}) + 2^n 2^{-n} = O(n),$$

so the function $f$ satisfies the condition (7.2), but its square $f^2$ does not:

$$\int_I f^2(w)\mu(w) = n^2(1 - 2^{-n}) + 2^{2n} 2^{-n} = O(2^n).$$

The following definition gives a wider class of polynomial on $\mu$-average functions which is closed under addition, multiplication, and multiplication by a scalar.

**Definition 7.2.3.** [Spherical definition] A function $f : I \to \mathbb{R}^+$ is polynomial on $\mu$-average on spheres if there exists an $\varepsilon > 0$ such that

$$\int_{I_n} f^\varepsilon(w) \mu_n(w) = O(n). \tag{7.3}$$

It is convenient, sometimes, to write the condition (7.3) in the following form (using the fact that $s(w) = n$ for $w \in I_n$):

$$\int_{I_n} \frac{f^\varepsilon(w)}{s(w)} \mu_n(w) = O(1).$$

The following proposition shows that every function which is expected polynomial on spheres satisfies the spherical definition of polynomial on $\mu$ average.

**Proposition 7.2.4.** *Let $\{\mu_n\}$ be an ensemble of spherical distributions. If a function $f : I \to \mathbb{R}^+$ is an expected polynomial on spheres relative to $\{\mu_n\}$, then it is a polynomial on $\mu$-average on spheres (satisfies Definition 7.2.3).*

*Proof.* Suppose for some $k \geq 1$,

$$\int_{I_n} f(w) \mu_n(w) \leq cn^k.$$

Put $\varepsilon = \frac{1}{k}$ and denote $S_n = \{w \in I_n \mid f^\varepsilon(w) \leq s(w)\}$. Then

$$\int_{I_n} \frac{f(w)^\varepsilon}{s(w)} \mu_n(w) = \int_{S_n} \frac{f(w)^\varepsilon}{s(w)} \mu_n(w) + \int_{\bar{S}_n} \frac{f(w)^\varepsilon}{s(w)} \mu_n(w)$$

$$\leq 1 + \int_{I_n} \left( \frac{f(w)^\varepsilon}{s(w)} \right)^k \mu_n(w) \leq 1 + \int_{I_n} \frac{f(w)}{n^k} \mu_n(w) \leq 1 + c.$$

Therefore,

$$\int_I \frac{f(w)^\varepsilon}{s(w)} \mu(w) = \Sigma_n \left( \mu(I_n) \int_{I_n} \frac{f(w)^\varepsilon}{s(w)} \mu_n(w) \right) \leq (1+c) \Sigma_n \mu(I_n) = 1 + c.$$

Since $s(w) = n$ for $w \in I_n$ one has

$$\int_{I_n} f^\varepsilon(w) \mu_n(w) \leq (1+c)n$$

so $f$ is polynomial on $\mu$-average on spheres. $\qquad\square$

To relax the condition (7.3) rewrite it first in the form

$$\int_{I_n} f^\varepsilon(w)s(w)^{-1}\mu_n(w) = O(1).$$

Recall now, that $\mu_n(w) = \mu(w)/\mu(I_n)$, hence

$$\int_{I_n} f^\varepsilon(w)s(w)^{-1}\mu(w) = O(1)\mu(I_n).$$

Therefore, for some constant $C > 0$,

$$\int_I f^\varepsilon(w)s(w)^{-1}\mu(w) \le \Sigma_n C\mu(I_n) = C. \qquad (7.4)$$

The condition (7.4) describes a larger class of functions which are intuitively polynomial on $\mu$-average. This condition, as well as the definition below of polynomial on average functions, is due to Levin [93].

**Definition 7.2.5.** [Levin's definition] A function $f : I \to \mathbb{R}^+$ is polynomial on $\mu$-average if there exists $\varepsilon > 0$ such that

$$\int_I (f(w))^\varepsilon s(w)^{-1}\mu(w) < \infty.$$

It is convenient to reformulate this definition in the following equivalent form (see [7], [58]):

**Definition 7.2.6.** A function $f : I \to \mathbb{R}^+$ is linear on $\mu$-average if

$$\int_I f(w)s(w)^{-1}\mu(w) < \infty,$$

and $f$ is polynomial on $\mu$-average if $f \le p(l)$ for some linear on $\mu$-average function $l : I \to \mathbb{R}^+$ and a polynomial $p$.

As we have mentioned already, the class of polynomial on $\mu$-average functions described in Levin's definition contains all functions which are average polynomials on spheres (the Spherical Definition 7.2.3). It is not hard to see also that this class is closed under addition, multiplication, and multiplication by a scalar. From now on, by polynomial on $\mu$-average functions we refer to functions from Levin's definition.

We note that, in general, not every function which is polynomial on $\mu$-average is also an average polynomial on spheres. However, if the distribution $\mu$ satisfies some sufficiently strong non-flatness conditions, then the statement holds.

**Proposition 7.2.7.** [58] *Let $\mu$ be an atomic distribution on $I$ and $\{\mu_n\}$ the induced ensemble of spherical distributions. If there exists a polynomial $p(n)$ such that for every $n$ either $\mu(I_n) = 0$ or $\mu(I_n) \geq \frac{1}{p(n)}$, then every function $f : I \to \mathbb{R}^+$ which is polynomial on $\mu$-average is also an average polynomial on spheres (relative to the ensemble $\{\mu_n\}$).*

To formulate a useful criterion of polynomial on average functions we need one more notion. A function $f$ is a *rarity function* if the integral

$$\int_I f(w)\mu(w)$$

(the expected value of $f$) converges.

**Proposition 7.2.8.** [58] *A function $f : I \to \mathbb{R}^+$ is polynomial on $\mu$-average if and only if there exists a rarity function $h : I \to \mathbb{R}^+$ and a polynomial $p(x)$ such that for every $w \in I$,*

$$f(w) \leq p(s(w), h(w)).$$

As a corollary we get another sufficient condition for a function $f$ to be polynomial on $\mu$-average, that is used sometimes as an alternative way to introduce polynomial on average functions.

**Corollary 7.2.9.** *If for some polynomial $p(x)$,*

$$\int_I \frac{f(w)}{p(s(w))}\mu(w) < \infty,$$

*then the function $f$ is polynomial on $\mu$-average.*

*Proof.* Indeed, in this case

$$\int_I \frac{f(w)}{s(w)p(s(w))}\mu(w) \leq \int_I \frac{f(w)}{p(s(w))}\mu(w) < \infty,$$

so the function $l(w) = \frac{f(w)}{p(s(w))}$ is linear on $\mu$-average. Hence $f(w) = l(w)p(s(w))$ is polynomial on $\mu$-average. $\square$

One more definition of polynomial on $\mu$-average functions was formally introduced by Impagliazzo in [68]. This is a volume analog of the spherical definition above, but contrary to the spherical case it is equivalent to Levin's definition. Impagliazzo's definition is very natural and allows one to work only with distributions on finite sets – the balls $B_n$ of $I$.

**Definition 7.2.10.** [Volume definition] *Let $\{\mu_n\}$ be an ensemble of volume distributions on balls $\{B_n\}$ of $I$. A function $f : I \to \mathbb{R}$ is polynomial on average with respect to $\{\mu_n\}$ if there exists an $\varepsilon > 0$ such that*

$$\int_{B_n} f^\varepsilon(x)\mu_n(x) = O(n).$$

It has been noticed in Section 7.1.2 that an atomic measure $\mu$ on $I$ induces an ensemble of atomic volume distributions $\{\mu_n\}$ on balls $B_n$, and vice versa, every ensemble of atomic distributions $\{\mu_n\}$ on balls $B_n$ such that $\mu_{n-1}$ is the distribution induced by $\mu_n$ on $B_{n-1}$, gives rise to an atomic distribution $\mu'$ on $I$ which induces $\{\mu_n\}$. In this case one has two different definitions of a function $f : I \to \mathbb{R}^+$ to be polynomial on average relative to $\mu$ and $\{\mu_n\}$. However, the following result shows that these definitions give the same class of polynomial on average functions.

**Proposition 7.2.11.** [68] *Let $\mu$ be a distribution on $I$ and $\{\mu_n\}$ the corresponding induced ensemble of volume distributions. Then a function $f : I \to \mathbb{R}^+$ is polynomial on $\mu$-average if and only if it is polynomial on average relative to the ensemble $\{\mu_n\}$.*

*Proof.* Suppose $f : I \to \mathbb{R}^+$ is polynomial on $\mu$-average, so there is $\varepsilon > 0$ such that

$$\int_I f(w)^\varepsilon s(w)^{-1}\mu(w) < \infty.$$

Then

$$\int_{B_n} f(w)^\varepsilon \mu_n(w) \le \int_{B_n} \frac{n}{s(w)} f(w)^\varepsilon \frac{\mu(w)}{\mu(B_n)}$$

$$\le \frac{n}{\mu(B_m)} \int_{B_n} \frac{1}{s(w)} f(w)^\varepsilon \mu(w) = O(n),$$

where $B_m$ is the ball of minimal radius with $\mu(B_m) \ne 0$ (such a ball always exists). Hence, $f$ is polynomial on average relative to $\{\mu_n\}$.

Conversely, assume now that $f$ is polynomial on average relative to $\{\mu_n\}$, so for some $\varepsilon > 0$,

$$\int_{B_n} f(x)^\varepsilon \mu_n(x) = O(n).$$

Put $S = \{w \in I \mid f(w)^{\frac{\varepsilon}{3}} \le s(w)\}$. Then

$$\int_I f(w)^{\frac{\varepsilon}{3}} s(w)^{-1}\mu(w) = \int_S f(w)^{\frac{\varepsilon}{3}} s(w)^{-1}\mu(w) + \int_{\overline{S}} f(w)^{\frac{\varepsilon}{3}} s(w)^{-1}\mu(w)$$

$$\le \int_I \mu(w) + \int_{\overline{S}} \frac{f(w)^\varepsilon}{f(w)^{\frac{2\varepsilon}{3}} s(w)}\mu(w) \le 1 + \Sigma_n \int_{I_n} \frac{f(w)^\varepsilon}{n^3}\mu(w)$$

$$\le 1 + \Sigma_n \frac{1}{n^3} \int_{B_n} f(w)^\varepsilon \mu(w) \le 1 + \Sigma_n \frac{1}{n^3} \int_{B_n} f(w)^\varepsilon \mu_n(w)$$

$$= 1 + \Sigma_n \frac{O(n)}{n^3} < \infty$$

and the leftmost integral above converges, as required.   $\square$

The results of this section show that the class of polynomial on average functions is very robust and, it seems, it contains all the functions that naturally look like "polynomial on average".

## 7.2.2 Average case behavior of functions

Definition 7.2.7 allows one to introduce general complexity classes on average.

**Definition 7.2.12.** Let $f : I \to \mathbb{R}$ and $t : \mathbb{R} \to \mathbb{R}$ be two functions. Then $f$ is $t$ on $\mu$-average if $f(w) = t(l(x))$ for some linear on $\mu$-average function $l$.

This definition can be reformulated in a form similar to Levin's Definition 7.2.6. To this end suppose for simplicity that a function $t : \mathbb{R}^+ \to \mathbb{R}^+$ is continuous, unbounded and monotone increasing, or $t : \mathbb{N}^+ \to \mathbb{N}^+$ and it is injective and unbounded. In the latter case, define for $x \in \mathbb{N}$,

$$t^{-1}(x) = \min\{y \mid t(y) \geq x\}.$$

Then it is not hard to show that a function $f : I \to \mathbb{R}$ is $t$ on $\mu$-average if and only if

$$\int_I t^{-1}(f(x))s(x)^{-1}\mu(x) < \infty.$$

For a function $t$ one can consider the class $Ave_\mu(t)$ of functions $f : I \to \mathbb{R}$ which are $t$ on $\mu$-average. Varying the function $t$ one may expect to get a hierarchy of functions on $\mu$-average. However, it is not always the case.

**Example 7.2.13.** If a function $f : I \to \mathbb{R}$ is $a^x$ on $\mu$-average for some $a > 1$, then $f$ is $b^x$ on $\mu$-average for any $b > 1$.

*Proof.* Indeed,

$$\int_I \frac{\log_b(f(x))}{s(x)}\mu(x) = \frac{1}{\log_b a}\int_I \frac{\log_a(f(x))}{s(x)}\mu(x) < \infty,$$

so $f$ is $b^x$ on $\mu$-average. □

We will have more to say on the hierarchy of time complexity of algorithmic problems in Section 7.2.3.

## 7.2.3 Average case complexity of algorithms

In this section we apply the average case analysis of functions to average case behavior of algorithms and time complexity of algorithmic problems.

Let $\mathcal{D}$ be a stratified distributional algorithmic problem, $I = I_{\mathcal{D}}$ the set of instances of $\mathcal{D}$ equipped with a size function $s = s_{\mathcal{D}} : I \to \mathbb{N}$ and an atomic probability distribution $\mu = \mu_{\mathcal{D}}$.

If $\mathcal{A}$ is a total decision algorithm for $\mathcal{D}$, then one can estimate overall efficiency of the algorithm $\mathcal{A}$ by its *expected running time*

$$\int_I T_{\mathcal{A}}(w)\mu(w).$$

Another way to characterize the "average" behavior of $\mathcal{A}$ relative to the size of inputs comes from the average case analysis of the time function $T_{\mathcal{A}}$.

An algorithm $\mathcal{A}$ has *polynomial time on $\mu$-average* if $T_{\mathcal{A}} : I \to \mathbb{N}$ is polynomial on $\mu$-average function. We say that $\mathcal{A}$ has *polynomial time upper bound on $\mu$-average* if $T_{\mathcal{A}}(x) \leq f(x)$ for some polynomial on $\mu$-average function $f : I \to \mathbb{N}$. Similarly, $\mathcal{A}$ has *time upper bound $t(x)$ on $\mu$-average* if $T_{\mathcal{A}}$ has an upper bound which is $t$ on $\mu$-average.

The notions above allow one to introduce complexity classes of algorithmic problems.

**Definition 7.2.14.** Let $\mathcal{D}$ be a stratified distributional problem. Then:

1) $\mathcal{D}$ is *decidable in polynomial time on average (AvePTime)* if there exists a polynomial time on $\mu$-average decision algorithm $\mathcal{A}$ for $\mathcal{D}$. The class of stratified distributional problems decidable in *AvePTime* is denoted by **AvePTime** (or simply by **AveP**).

2) $\mathcal{D}$ is *decidable in $t$ time on average (AveTime(t))* for some time bound $t$ if there exists a decision algorithm for $\mathcal{D}$ with a time upper bound $t$ on $\mu$-average. The class of stratified distributional problems decidable in *AveTime(t)* is denoted by **AveTime(t)**.

### 7.2.4   Average case vs worst case

In this section we compare the average case complexity to the worst case one. To do this we briefly discuss what kind of "hardness" the average case complexity does measure, indeed. We refer to Gurevich's paper [60] for a detailed discussion of these issues.

The main outcome of the development of average case complexity is that this type of complexity gives a robust mathematical framework and provides one with a much more balanced view-point on the computational hardness of algorithmic problems than the worst case complexity. It turns out for example that many algorithmic problems that are hard in the worst case are easy on average.

Consider, for instance, the Hamiltonian Circuit Problem (HCP). Recall that a *Hamiltonian circuit* in a graph is a closed path that contains every vertex exactly once. HCP asks for a given finite graph $\Gamma$ to find a Hamiltonian circuit in $\Gamma$. It is known that this problem is **NP**-complete (see, for example, [44]). However, Gurevich and Shelah showed in [61] that there is a decision algorithm for the HCP which is linear time on average. More precisely, if one has a distribution on finite graphs such that any two vertices $u, v$ in a given random finite graph $\Gamma$ are connected by an edge with a fixed uniform probability $\alpha$, $0 < \alpha < 1$, then there is an algorithm that solves the problem in linear time on average.

**Corollary 7.2.15.** *There are **NP***-complete problems which are polynomial on average with respect to some natural distributions.*

### 7.2.5 Average case behavior as a trade-off

The concept of average case behavior of functions may look sometimes counterintuitive. In this section we discuss what kind of knowledge the average case analysis of functions or algorithms brings to the subject. We argue that the average case analysis is a wrong tool when one tries to describe the "typical behavior" of a function, say its behavior on "most" or "typical" inputs. We follow Gurevich [58, 60] and Impagliazzo [68] in the discussion on the essence of the concept of a polynomial on average function. At the end of the section we give a convenient tool to get general upper bounds on average.

To indicate the typical problems with the average case behavior we give two examples.

**Example 7.2.16.** Let $I = \{0,1\}^*$ be the set of all words in the alphabet $\{0,1\}$. Define the size of a word $w \in I$ to be equal to the length $|w|$ of the word, and let the distribution $\mu$ on $I$ be such that its restriction on a sphere $I_n$ is uniform for every $n$. Denote by $S_n$ a subset of $I_n$ of measure $\frac{2^n-1}{2^n}$, so $S_n$ has precisely $2^n - 1$ elements. Now we define a function $l : I \to \mathbb{N}$ by its values on each sphere $I_n$ as follows:

$$l(w) = \begin{cases} 0, & \text{if } w \in S_n, \\ 2^n, & \text{if } w \notin S_n. \end{cases}$$

Then

$$\int_I l(w)|w|^{-1}\mu(w) < \infty$$

so the function $l$ is linear on $\mu$-average. Now we define a function $f : I \to \mathbb{N}$ by its values on each sphere $I_n$ by:

$$f(w) = \begin{cases} 1, & \text{if } w \in S_n, \\ 2^{2^n}, & \text{if } w \notin S_n. \end{cases}$$

Clearly, $f(w) = 2^{l(w)}$, so $f$ is exponential on $\mu$-average. However, $f(w) = 1$ on most words $w \in I$.

Another type of a strange behavior of functions polynomial on average is described in the following example.

**Example 7.2.17.** Let $I = \{0,1\}^*$. Define the size of a word $w \in I$ to be equal to the length $|w|$ of the word, and let the distribution $\mu$ on $I$ be given by

$$\mu(w) = 2^{-2|w|-1}.$$

If $f : I \to \mathbb{N}$ is a function such that $f(w) = 2^{|w|}$. Then

$$\sum_{|w|\neq 0} \frac{f^\varepsilon(w)}{|w|}\mu(w) = \sum_{n=1}^{\infty}\sum_{|w|=n} \frac{2^{n\varepsilon}}{2^{2n+1}} = \sum_{n=1}^{\infty} 2^{n\varepsilon-n-1}$$

which converges for any $\varepsilon < 1$. Hence $f$ is polynomial on $\mu$-average. On the other hand, the function $f(x)$ is clearly "exponential on most inputs".

To properly understand the examples above one has to focus on a general question:

> What kind of knowledge does the average case analysis of behavior of functions bring into the subject?

In [60] Gurevich explains, in terms of a Challenger-Solver game, what kind of hardness the concept of the average case complexity is aiming for. First of all, the examples above show that it is not the cost for the Solver to win the game on the most typical inputs, or even expected cost, rather, it captures the *trade-off* between a measure of difficulty and the fraction of hard instances of the problem. Thus, to have polynomial on average time an algorithm should have only a sub-polynomial fraction of inputs that require superpolynomial time to compute.

The following definition (due to Impagliazzo [68]) is a rigorous attempt to grasp this type of trade-off to stay in the class of polynomial on $\mu$-average functions.

Let $\mathcal{D}$ be a stratified distributional algorithmic problem, $I = I_{\mathcal{D}}$, $s = s_{\mathcal{D}}$, and $\mu = \{\mu_n\}$ be an ensemble of volume distributions for $I$. A partial algorithm $\mathcal{A}$ solves $\mathcal{D}$ with *benign faults* if for every input $w \in I$ it either outputs the correct answer or it stops and says "?" ("Do not know"). A polynomial time *benign algorithm scheme* for $\mathcal{D}$ is an algorithm $\mathcal{A}(w, \delta)$ such that for any input $w \in I$ and $\delta \in \mathbb{N}$ the algorithm $\mathcal{A}$ does the following:

- there is a polynomial $p(x, y)$ such that $\mathcal{A}$ stops on $(w, \delta)$ in time $p(s(w), \delta)$;

- $\mathcal{A}$ solves $\mathcal{D}$ on input $w$ with benign faults;

- for any $\delta \in \mathbb{N}$ and $n \in \mathbb{N}$, $\mu_n(\{w \in B_n \mid A(w, \delta) =?\}) \leq \frac{1}{\delta}$.

The following results show equivalence of polynomial on average problems and existence of polynomial time benign algorithm schemes.

**Lemma 7.2.18 (Impagliazzo [68]).** *A stratified distributional algorithmic problem $\mathcal{D}$ with an ensemble of volume distributions $\mu = \{\mu_n\}$ is polynomial on $\mu$-average if and only if $\mathcal{D}$ has a polynomial time benign decision algorithm scheme $\mathcal{A}(w, \delta)$.*

Now we can properly analyze the examples above.

Example 7.2.17 reveals "traces" of worst case behavior in average case complexity. Indeed, in this example the balance between the time complexity of an algorithm on the worst case inputs and the measure of the fraction of these inputs is broken, i.e., the fraction of hard instances is small but the time required by the algorithm to finish computation on these instances is very large (compare to the size of the fraction), therefore the average case complexity does not reflect behavior of the algorithm on most inputs.

In Example 7.2.18, the balance between the fraction of hard inputs and the measure of this fraction is almost perfect, therefore the function "looks polynomial" on average, which is, of course, not the case.

The cause of this discrepancy is reflected in the notion of a flat distribution (see e.g., [58]). One may try to avoid such distributions altogether, but they appear quite naturally in many typical situations (see [58, 60] and [17]), so it is difficult to ignore them. On the other hand, if one puts into the game a natural non-flatness condition, say as in Proposition 7.2.8, then polynomial on average functions become polynomial on average on spheres, that makes the class of polynomial on average functions smaller. In any case, Example 7.2.18 is troubling.

To deal with the trade-off in the general case we introduce the following definitions. A function $h : \mathbb{Z}^+ \to \mathbb{R}^+$ is $\mu$-*balanced* on a subset $K \subseteq I$ if

$$\Sigma_n h(n) n^{-1} \mu(K \cap I_n) < \infty.$$

A function $U : \mathbb{R}^+ \to \mathbb{R}^+$ is an $s$-*upper bound* for $f : I \to \mathbb{R}^+$ on a subset $K \subseteq I$ if $f(w) \leq U(s(w))$ for every $w \in K$. Finally, $f$ is $(s, \mu)$-*balanced on* $K$ if there exists an $s$-upper bound $U$ for $f$ on $K$ which is $\mu$-balanced on $K$, i.e.,

$$\Sigma_n U(n) n^{-1} \mu(K \cap I_n) < \infty.$$

**Proposition 7.2.19 (Balance test).** *Let* $f : I \to \mathbb{R}^+$ *and* $f_1 : \mathbb{R}^+ \to \mathbb{R}^+$ *be two functions. Suppose* $f_1$ *is a bijection. If there exists a subset* $K \subseteq I$ *such that*

1) $f$ *is* $f_1$ *on* $\mu$-*average on* $K$ *(in the* $\mu$-*induced measure* $\mu_K$ *on* $K$*),*

2) $f_1^{-1}(f)$ *is* $(s, \mu)$-*balanced on* $I - K$,

*then* $f$ *is* $f_1$ *on* $\mu$-*average.*

*Proof.* To show that $f$ is $f_1$ on $\mu$ average it suffices to prove that the integral

$$\int_I \frac{f_1^{-1}(f(w))}{s(w)} \mu(w)$$

converges. To see this consider

$$\int_I \frac{f_1^{-1}(f(w))}{s(w)} \mu(w) = \int_K \frac{f_1^{-1}(f(w))}{s(w)} \mu(w) + \int_{I-K} \frac{f_1^{-1}(f(w))}{s(w)} \mu(w)$$

$$\leq \mu(K) \int_K \frac{f_1^{-1}(f(w))}{s(w)} \mu_K(w) + \int_{I-K} \frac{U(s(w))}{s(w)} \mu(w)$$

(here $U$ is an $s$-upper bound on $I - K$ for $f_1^{-1}(f)$ which is $(s, \mu)$-balanced on $I - K$)

$$\leq \mu(K) \int_K \frac{f_1^{-1}(f(w))}{s(w)} \mu_K(w) + \Sigma_n \frac{U(n)}{n} \mu((I - K) \cap I_n) < \infty.$$

This shows that $T_A$ runs within the bound $f_1$ on $\mu$-average, as claimed. $\square$

**Remark 7.2.20.** The condition on $f_1$ to be a bijection is introduced, just to ensure that $f_1^{-1}(f(w))$ always exists. One can easily replace it, if needed, with a more relaxed one. In the case of algorithms their time functions are integer functions of the type $f : \mathbb{N} \to \mathbb{N}$, in which case it suffices (to get a result similar to the Balance test) to assume that $f_1 : \mathbb{Z}^+ \to \mathbb{Z}^+$ is unbounded and define $f_1^{-1}(n) = \min\{m \mid f_1(m) \geq n\}$.

An adaptation of the argument above to the time functions of algorithms gives a robust test for an algorithm to run within a given upper bound on $\mu$-average. Below we assume that $f_1 : \mathbb{Z}^+ \to \mathbb{Z}^+$ is unbounded and $f^{-1}$ is defined as above.

**Proposition 7.2.21 (Average running time test).** *Let $\mathcal{A}$ be an algorithm with a set of inputs $I$. If there exists a subset $K \subseteq I$ such that*

1) *$\mathcal{A}$ runs on $K$ within an upper bound $f_1$ on $\mu$-average (in the $\mu$-induced measure on $K$),*

2) *there exists an upper bound $f$ for $\mathcal{A}$ on $I - K$ such that $f_1^{-1}(f)$ is $(s, \mu)$-balanced on $I - K$,*

*then $\mathcal{A}$ runs on $I$ within the time bound $f_1$ on $\mu$-average.*

*Proof.* Observe, that $T_{\mathcal{A}}(w) \leq f_1(s(w))$ if $w \in K$ and $T_{\mathcal{A}}(w) \leq f(s(w))$ otherwise. We claim that the algorithm $\mathcal{A}$ runs in time $f_1$ on $\mu$-average. To see this consider

$$\int_I \frac{f_1^{-1}(T_{\mathcal{A}}(w))}{s(w)} \mu(w) = \int_K \frac{f_1^{-1}(T_{\mathcal{A}}(w))}{s(w)} \mu(w) + \int_{I-K} \frac{f_1^{-1}(T_{\mathcal{A}}(w))}{s(w)} \mu(w)$$

$$\leq \int_K \frac{f_1^{-1}(f_1(s(w)))}{s(w)} \mu(w) + \int_{I-K} \frac{f_1^{-1}(f(w))}{s(w)} \mu(w)$$

$$\leq \mu(K) + \int_{I-K} \frac{U(s(w))}{s(w)} \mu(w) < \infty.$$

This shows that $T_{\mathcal{A}}$ runs within the bound $f_1$ on $\mu$-average, as claimed. □

**Remark 7.2.22.** It is easy to show particular examples of $K$, $f$, and $f_1$ to satisfy the conditions of Proposition 7.2.22. Some of them are given in Corollary 8.2.5 of Section 8.2.3.

## 7.2.6 Deficiency of average case complexity

In this section we argue that despite the fact that average case complexity embodies a much more practical viewpoint on complexity of algorithms (than the worst case complexity) it does not, in general, provide the type of analysis of algorithms which is required in many applications where the focus is on behavior of algorithms on most or typical inputs. For example, in cryptography the average case analysis does not play much of a role, since (see Example 7.2.17) an algorithm, that lies at the heart of the algorithmic security of a cryptosystem, can be of exponential time on average (so presumably hard), and have linear time on most inputs. Another typical example is Dantzig's algorithm, mentioned above, for linear programming, which is much more robust in applications than existing polynomial time (in the worst case!) algorithms. The main reason for the inadequate results of the average case analysis in many situations is, of course, that there are natural algorithms which have high (say, exponential) average case complexity, but still are very fast on most inputs; and conversely, there are obviously exponential time on most input functions which are polynomial on average relative to some quite natural distributions. In fact, it seems, it is not easy to come up with any real application which would require the average case analysis of algorithms, rather then the worst case analysis, or the analysis on the "most inputs".

One of the ways to introduce a complexity class that would reflect the behavior of algorithms on most inputs comes from statistics. The idea is to use the median time of the Solver, since the median better represents typical behavior. However, it is not clear at all how robust such a theory would be, and there are indications that difficulties similar to the average case naturally occur.

Our main point in this discussion is that quite often it would be convenient and more practical to work with another type of complexity which focuses on generic inputs of algorithms, ignoring sparse sets of "non-typical" inputs. This type of complexity, coined as *generic case complexity*, has been emerging in asymptotic mathematics for some time now, and it is quite close, in spirit, to the above mentioned results of Vershik and Sporyshev [140], and Smale [130] in their analysis of the simplex algorithm for linear programming problems.

# Chapter 8

# Generic Case Complexity

## 8.1 Generic Complexity

In this section we discuss a general notion of generic complexity of algorithms and algorithmic problems. Our exposition here is along the lines of [48].

### 8.1.1 Generic sets

We start by introducing some necessary notation.

Let $I$ be a set. Denote by $\mathcal{P}(I)$ the subset algebra of $I$, i.e., the set of all subsets of $M$ with operations of union, intersection and complementation.

A *pseudo-measure* on $I$ is a real-valued nonnegative function $\mu : \mathcal{A} \to \mathbf{R}^+$ defined on a subset $\mathcal{A} \subset \mathcal{P}(I)$ such that

1) $\mathcal{A}$ contains $M$ and is closed under disjoint union and complementation;

2) $\mu(I) = 1$ and for any disjoint subsets $A, B \in \mathcal{A}$,

$$\mu(A \cup B) = \mu(A) + \mu(B).$$

In particular, $\mu(\overline{A}) = 1 - \mu(A)$.

If $\mathcal{A}$ is a subalgebra of $\mathcal{P}(I)$, then $\mu$ is a measure. We make a point to consider pseudo-measures here since the asymptotic density function (see Section 8.1.2), which is one of the fundamental tools in asymptotic mathematics, in general, is only a pseudo-measure, not a measure.

A pseudo-measure $\mu$ is called *atomic* if $\mu(Q)$ is defined for any finite subset $Q$ of $I$.

**Definition 8.1.1.** Let $\mu$ be a pseudo-measure on $I$. We say that a subset $Q \subseteq I$ is *generic* if $\mu(Q) = 1$ and *negligible* if $\mu(Q) = 0$.

The following lemma is easy.

**Lemma 8.1.2.** *Let $I$ be a set equipped with a pseudo-measure $\mu$. The following holds for arbitrary subsets $S, T$ of $I$:*

1) *$S$ is generic if and only if its complement $\overline{S}$ is negligible;*

2) *If $S$ is generic and $S \subseteq T$, then $T$ is generic;*

3) *Finite unions and intersections of generic (negligible) sets is generic (negligible);*

4) *If $S$ is generic and $T$ is negligible, then $S - T$ is generic;*

5) *The set $\mathcal{B}$ of all generic and negligible sets forms an algebra of subsets of $I$;*

6) *$\mu : \mathcal{B} \to \mathbf{R}^+$ is a measure on $I$.*

If the set $I$ is countable and $\mu$ is an atomic probabilistic measure on $I$ with no elements of probability zero (for example, Exponential or Cauchy distributions from Section 7.1.1), then $I$ is the only generic set in $I$. In this situation the notions of generic and negligible sets are not very interesting. However, if the set $I$ has a natural stratification, then the measure $\mu$ gives rise to a very useful pseudo-measure on $I$ – the asymptotic density $\rho_\mu$ relative to $\mu$. We discuss this in the next section.

## 8.1.2  Asymptotic density

Let $I$ be a set with a size function $s : I \to \mathbb{N}^+$ (or $s : I \to \mathbb{R}^+$). Let $\mu = \{\mu_n\}$ be an ensemble of spherical distributions for $I$ (so $\mu_n$ is a distribution on the sphere $I_n$).

For a set $R \subseteq I$ one can introduce the *spherical asymptotic density* $\rho_\mu$, with respect to the ensemble $\mu$, as the following limit (if it exists):

$$\rho_\mu(R) = \lim_{n \to \infty} \mu_n(R \cap I_n).$$

The most typical example of an asymptotic density arises when all the spheres $I_n$ are finite.

**Example 8.1.3.** Suppose the spheres $I_n$ are finite for every $n \in \mathbb{N}$. Denote by $\mu_n$ the uniform distribution on $I_n$. Then for a subset $R \subseteq I$,

$$\mu_n(R) = \frac{|R \cap I_n|}{|I_n|}$$

is just the frequency of occurrence of elements from $R$ in the sphere $I_n$. In this case the spherical asymptotic density $\rho_\mu(R)$ is just the standard *uniform spherical asymptotic density* of $R$ (see Section 3.2.3). Usually, we denote $\rho_\mu(R)$ by $\rho(R)$, and $\mu_n(R)$ by $\rho_n(R)$.

The following example shows that for some sets $R$ the asymptotic density $\rho(R)$ does not exist.

**Example 8.1.4.** Let $I = X^*$ be the set of all words in a finite alphabet $X = \{x_1, \ldots, x_m\}$ and $R$ the subset of all words of even length. Then $\rho_n(R) = 1$ for even $n$ and $\rho_n(R) = 0$ for odd $n$, so the limit $\rho(R)$ does not exist.

One way to guarantee that the asymptotic density $\rho(R)$ always exists is to replace $\lim_{n \to \infty} \rho_n(R)$ with the $\limsup_{n \to \infty} \rho_n(R)$. In the sequel, we mostly use the standard limit (since the results are stronger in this case), and only occasionally use the upper limit (when the standard limit does not exist).

Another way to make the asymptotic density smoother is to replace spheres $I_n$ in the definition above with the balls $B_n = B_n(I)$. If $\nu = \{\nu_n\}$ is an ensemble of volume distributions for $I$, then the *volume* (or *disc*) asymptotic density $\rho_\nu^*$ relative to $\nu$ is defined for a subset $R \subseteq I$ as the limit (if it exists):

$$\rho_\nu^*(R) = \lim_{n \to \infty} \nu_n(R).$$

The following lemma is obvious.

**Lemma 8.1.5.** *The spherical and volume asymptotic densities on $I$ are pseudo-measures on $I$.*

Now we consider another typical way to define asymptotic densities of sets. Let, as above, $I$ be a set with a size function $s : I \to \mathbb{N}$. Suppose $\mu$ is a probability distribution on $I$ such that all the spheres $I_n$ are measurable subsets of $I$. Denote by $\mu_n$ the probability distribution on $I_n$ induced by $\mu$. In this case if $R$ is a measurable subset of $I$, then the frequencies $\mu_n(R)$ are well defined, so it makes sense to consider the spherical asymptotic density $\rho_\mu(R)$ or $R$. Observe, that in this case the balls $B_n$ are also measurable, as finite disjoint unions of $n$ spheres $B_n(I) = I_1 \cup \ldots \cup I_n$, so the volume frequencies are also defined for $R$, and the notion of the volume asymptotic density also makes sense. In particular, the probability distributions defined in Section 7.1.1 give rise to interesting asymptotic densities with nontrivial generic and negligible sets, as was mentioned before.

**Remark 8.1.6.** If the probability distribution $\mu$ on $I$ is size-invariant, i.e., for any $u, v \in I, \mu(u) = \mu(v)$ provided $s(u) = s(v)$, then it induces the uniform distribution on $I_n$, so the spherical asymptotic density with respect to $\mu$ on $I$ is equal to the standard uniform spherical asymptotic density on $I$. The same is true for volume size-invariant distributions.

At the end of this section we prove a lemma which shows that the standard uniform spherical and volume asymptotic densities are equal in some natural situations.

**Lemma 8.1.7.** *Let all the spheres $I_n$ be finite and nonempty. If the standard spherical density (as in Example 8.1.3) $\rho(R)$ exists for a subset $R$ of $I$, then the standard volume density $\rho^*(R)$ also exists and $\rho^*(R) = \rho(R)$.*

*Proof.* Set $x_n = |R \cap B_n|$ and $y_n = |B_n|$. Then $y_n < y_{n+1}$ and $\lim y_n = \infty$. By Stolz's theorem

$$\rho^*(R) = \lim_{n \to \infty} \frac{x_n}{y_n} = \lim_{n \to \infty} \frac{x_n - x_{n-1}}{y_n - y_{n-1}} = \lim_{n \to \infty} \frac{|R \cap I_n|}{|I_n|} = \rho(R),$$

as claimed.                                                                                    □

### 8.1.3   Convergence rates

Let $I$ be a set with a size function $s : I \to \mathbb{N}$. Suppose for each $n \in \mathbb{N}$ the sphere $I_n$ and the ball $B_n$ are equipped with probability distributions $\mu_n$ and $\mu_n^*$, correspondingly. The asymptotic densities $\rho_\mu$ and $\rho_\mu^*$ are defined and by Lemma 8.1.5 they are pseudo-measures. Hence the results from Section 8.1.1 apply and the notion of a generic and a negligible set relative to $\rho_\mu$ and $\rho_\mu^*$ are defined. We would like to point out here that these asymptotic densities not only allow one to distinguish between "large" (generic) and "small" (negligible) sets, but they provide a much finer method to describe asymptotic behavior of sets at "infinity" with respect to the given size function $s$. In this section we introduce the required machinery.

We start with the spherical asymptotic density $\rho_\mu$. One can introduce similar notions for the volume density $\rho_\mu^*$; we leave it to the reader.

Let $\mu = \{\mu_n \mid n \in \mathbb{N}\}$ be a fixed ensemble of spherical distributions for $I$.

**Definition 8.1.8.** Let $R$ be a subset of $I$ for which the asymptotic density $\rho_\mu(R)$ exists. A function $\delta_R : n \to \mu_n(R \cap I_n)$ is called a *frequency function* of $R$ in $I$ with respect to the spherical ensemble $\mu$. Its dual $\delta_{\bar R} : n \to \mu_n(\bar R \cap I_n)$ is called the *residual probability function* for $R$.

The density function $\delta_R$ may show how quickly the frequencies $\mu_n(R \cap I_n)$ converge to the asymptotic density $\rho(R)$ (if it exists), henceforth the convergence rate of $\delta_R$ gives a method to differentiate between various generic or negligible sets.

**Definition 8.1.9.** Let $R \subseteq I$ and $\delta_R$ be the frequency function of $R$. We say that $R$ has asymptotic density $\rho(R)$ with the convergence rate

1) of *degree* $n^k$ if $|\rho(R) - \delta_R(n)| \sim cn^{-k}$ for some constant $c$;

2) *faster than* $n^{-k}$ if $|\rho(R) - \delta_R(n)| = o(n^{-k})$;

3) *superpolynomial* if $|\rho(R) - \delta_R(n)| = o(n^{-k})$ for any natural $k$;

4) *subexponential* if $|\rho(R) - \delta_R(n)| = o(r^n)$ for every $r > 1$.

4) *exponential* if $|\rho(R) - \delta_R(n)| \sim c^n$ for some $0 < c < 1$.

Of course, one can introduce different degrees of exponential convergence, superexponential convergence, etc.

Now one can distinguish various generic (negligible) sets with respect to their convergence rates. Thus, we say that $R$ is *superpolynomially (exponentially)* generic if it is generic and its convergence rate is superpolynomial (exponential). Sometimes we refer to superpolynomial generic sets as *strongly* generic sets. Note that in the original papers [75, 76] the strongly generic sets meant exponentially generic sets, but it turned out that in most applications it is more convenient to have a special name for superpolynomial generic sets. We define *superpolynomial* and *exponentially negligible* sets as the complements of the corresponding generic sets.

**Example 8.1.10.** Let $I = X^*$ be the set of all words in a finite alphabet $X$ with the length of the word as the size function and with the standard uniform distribution on spheres $I_n$. If $R = wX^*$ is the cone generated by a given word $w \in I$, then $\rho(R) = |X|^{-|w|}$, so $R$ is neither generic nor negligible.

## 8.1.4 Generic complexity of algorithms and algorithmic problems

In this section we introduce generic complexity of algorithms and algorithmic problems. For simplicity we consider only spherical distributions here, leaving it to the reader to make obvious modifications in the case of volume distributions.

We start with a general notion of generic decidability of distributional algorithmic (decision or search) problems.

**Definition 8.1.11.** Let $\mathcal{D}$ be a distributional computational problem. A partial decision algorithm $\mathcal{A}$ for $\mathcal{D}$ *generically solves* the problem $\mathcal{D}$ if the halting set $H_{\mathcal{A}}$ of $\mathcal{A}$ is generic in $I = I_{\mathcal{D}}$ with respect to the given probability distribution $\mu = \mu_{\mathcal{D}}$ on $I$. In this case we say that $\mathcal{D}$ is *generically decidable*.

Note that a priori, we do not require that a generically decidable problem be decidable in the worst case.

To discuss generic complexity of algorithms and algorithmic problems one needs, as usual, to have a size function $s : I \to \mathbb{N}$ on the set of inputs $I = I_{\mathcal{D}}$. Let $\mu = \{\mu_n\}$ be an ensemble of spherical distributions on $I$ relative to the size function $s$. Let $\mathcal{A}$ be a partial decision algorithm for $\mathcal{D}$ with the halting set $H_{\mathcal{A}}$, and a partial time function $T_{\mathcal{A}} : I \to \mathbb{N}$ (see Section 3.3 for definitions). We assume here that if $\mathcal{A}$ does not halt on $x \in I$, then $T_{\mathcal{A}}(x) = \infty$.

**Definition 8.1.12.** Let $\mathcal{D}$ be a stratified distributional computational problem and $\mathcal{A}$ a partial decision algorithm for $\mathcal{D}$. A time function $f(n)$ is a *generic upper bound* for $\mathcal{A}$ if the set

$$H_{\mathcal{A},f} = \{w \in I \mid T_{\mathcal{A}}(w) \leq f(s(w))\}$$

is generic in $I$ with respect to the spherical asymptotic density $\rho_\mu$. In this case, the residual probability function

$$\mathcal{C}_{\mathcal{A},f}(n) = 1 - \rho_n(H_{\mathcal{A},f}) = 1 - \mu_n(H_{\mathcal{A},f} \cap I_n)$$

is termed the *control sequence* of the algorithm $\mathcal{A}$ relative to $f$.

Similarly, $f(n)$ is a *strongly generic (exponentially generic*, etc.) time upper bound for $\mathcal{A}$ if the set $H_{\mathcal{A},f}$ is strongly generic (has exponential convergence rate, etc.).

Now we are ready to define generic complexity classes of algorithmic problems.

**Definition 8.1.13.** We say that a stratified distributional decision problem $\mathcal{D}$ is

- *decidable generically in polynomial time* (or *GPtime decidable*) if there exists a decision algorithm $\mathcal{A}$ for $\mathcal{D}$ with a generic polynomial upper bound $p(n)$.

- *decidable strongly generically in polynomial time* (or *SGPtime decidable*) if there exists a decision algorithm $\mathcal{A}$ for $\mathcal{D}$ with a strongly generic polynomial upper bound $p(n)$.

In the situation above we say that $\mathcal{D}$ has *generic (strongly generic) time complexity* at most $p(n)$.

By **GenP** and **Gen$_{\mathbf{str}}$P** we denote the classes of problems decidable generically and, respectively, strongly generically, in polynomial time.

Similarly, one can introduce generic complexity classes for arbitrary time bounds. Furthermore, it is sometimes convenient to specify the density functions $\delta$ for the generic sets involved.

**Definition 8.1.14 (Generic Deterministic Time).** Let $f : \mathbb{N} \to \mathbb{R}$. By **Gen$_{\delta}$TIME**$(f)$ we denote the class of stratified distributional problems $(\mathcal{D}, \mu)$ such that there exists a partial decision algorithm $\mathcal{A}$ for $\mathcal{D}$ whose set $H_{\mathcal{A},f}$ is generic (relative to the spherical $\rho$, or volume $\rho^*$, asymptotic density with respect to $\mu$) in $I$ with the density function $\delta$.

## 8.1.5   Deficiency of the generic complexity

In this section we discuss deficiencies of the generic complexity. One obvious deficiency is common to all complexity classes of distributional algorithmic problems. It comes from a choice of measure: if the measure is unnatural, the results could be counterintuitive. Consider the following example.

**Example 8.1.15.** Let $I = \{0,1\}^*$. Define the size of a word $w \in I$ to be equal to the length $|w|$ of the word. Suppose for each $n$ one has a disjoint decomposition of the sphere $I_n$ into two subsets $I_n = I'_n \cup I''_n$ of size $2^{n-1}$. Put

$$I' = \cup_n I'_n, I'' = \cup_n I''_n$$

and define a function $f : I \to \mathbb{N}$ by

$$f(w) = \begin{cases} 0, & \text{if } w \in I', \\ 2^{|w|}, & \text{if } w \in I''. \end{cases}$$

Then one can construct two measures $\mu'$ and $\mu''$ (which are positive on nonempty words) such that $f$ is generically exponential with respect to the asymptotic density relative to $\mu'$ and generically constant with respect to the asymptotic density relative to $\mu''$. Indeed, define $\mu'$, such that $\mu'(I'_n) = \frac{2^{n+1}-1}{2^{n+1}}$ and $\mu'(I''_n) = \frac{1}{2^{n+1}}$. Then $I'$ is generic with respect to the asymptotic density $\rho_{\mu'}$ (so $I''$ is negligible). Similarly, one can define $\mu''$ such that $I''$ becomes generic (and $I'$ becomes negligible).

The only recipe here to avoid bad examples is to choose the initial measure that reflects the nature of the problem.

## 8.2 Generic versus average case complexity

In Section 8.2.1 below, the average case complexity is compared to generic case complexity. We claim that generic case complexity is a better tool for describing behavior of algorithms on most inputs, so it is more useful in some typical applications.

It turns out however that despite all the differences, generic- and average case complexities in many particular cases are quite close to each other.

In Section 8.2.2 we show that in many cases, if an algorithm is easy on average, then it is also easy generically. The opposite is not true, in fact there are exponential on average algorithms that are polynomial generically.

In Section 8.2.3 we give a robust test that shows when a generic upper time bound of an algorithm is also a time bound on average.

### 8.2.1 Comparing generic and average case complexities

In this section we discuss principal differences between the average and generic case complexities.

One obvious distinction is that in generic case complexity one can consider undecidable problems as well as decidable ones. Moreover, the approach is uniform to both types of problems. For example, in [62] the generic case complexity of the Halting Problem for Turing machines is studied.

Another difference is that generic case complexity aims at the typical behavior of the algorithm, i.e., behavior on most inputs, rather than on the expected time, or the trade-off between the fraction of hard inputs and the computation time of the algorithm on these inputs.

One really big advantage of generic case complexity is that for a given problem it is much easier to find (and prove) a fast generic algorithm, than to find (and prove!) a fast algorithm on average. We refer to Chapter 9 for support of this claim.

We would also like to note that generic complexity is much clearer conceptually than average case complexity. Indeed, if the measure is fixed, then generic case complexity describes precisely what it claims to describe, i.e., the behavior of a function on "most" inputs (i.e., on a generic set of inputs), while average case complexity measures an allusive "trade-off", which is not easy to formalize, let alone to measure.

Finally, we claim that what counts in most applications is precisely the generic behavior of algorithms, see the discussion in Section 7.2.6.

## 8.2.2   When average polynomial time implies generic polynomial time

We have seen examples (see Example 7.2.18) where polynomial on average functions are not generically polynomial. However, if we restrict the class of polynomial on average functions to the narrower class of functions which are polynomial on average on spheres (see Definition 7.2.1 in Section 7.2.1), then all functions in this class are generically polynomial. This shows that, perhaps, Levin's class of polynomial on average functions [93] is too wide.

Let $\mathcal{D}$ be a stratified distributional algorithmic problem with a set of instances $I$ equipped with a size function $s : I \to \mathbb{N}$ and an atomic distribution $\mu$. It has been noted in Section 7.1.2 that an atomic measure on $I$ induces an ensemble of atomic distributions $\{\mu_n\}$ on spheres $I_n$. Recall that a function $f : I \to \mathbb{R}^+$ is polynomial on $\mu$-average on spheres if there exists $k \geq 1$ such that

$$\int_{I_n} f^{\frac{1}{k}}(w)\mu_n(w) = O(n).$$

**Proposition 8.2.1.** *If a function $f : I \to \mathbb{R}^+$ is polynomial on $\mu$-average on spheres, then $f$ is generically polynomial relative to the asymptotic density $\rho_\mu$.*

*Proof.* If $f$ is an expected polynomial, then there exist a constant $c$ and $k \geq 1$ such that for any $n$,

$$\int_{I_n} f^{\frac{1}{k}}(w)\mu_n(w) \leq cn.$$

It follows that for any polynomial $q(n)$,

$$\mu_n\{x \in I_n \mid f^{\frac{1}{k}}(x) > q(n)cn\} \leq 1/q(n).$$

Now let $S(f, q, k) = \{x \in I \mid f(x) \geq (cq(s(x))s(x))^k\}$ be the set of those instances from $I$ on which $f(x)$ is not bounded by $(cq(s(x))s(x))^k$. Then

$$\mu_n(I_n \cap S(f, q, k)) = \mu_n\{x \in I_n \mid f^{\frac{1}{k}}(x) > q(n)cn\} \leq 1/q(n).$$

Therefore, the asymptotic density $\rho_\mu$ of $S(f, q, k)$ exists and is equal to 0. This shows that $f$ is generically bounded by the polynomial $(cq(n)n)^k$.     □

Proposition 8.2.1 gives a large class of polynomial on average functions that are generically polynomial.

**Corollary 8.2.2.** *Let $\mathcal{A}$ be a decision algorithm for the algorithmic problem $\mathcal{D}$. If the expected time of $\mathcal{A}$ (with respect to the spherical distributions) is bounded by a polynomial, then $\mathcal{A}$ GPtime decides $\mathcal{D}$. Moreover, increasing the generic time-complexity of $\mathcal{D}$ we can achieve any control sequence converging to $0$ polynomially fast.*

The result of Proposition 8.2.1 can be generalized to arbitrary time bounds $t(x)$ from Section 7.2.2, but in a slightly weaker form. Here, instead of a generic bound $t(x)$, one gets a generic bound $t(x \log^* x)$, where $\log^*$ is an extremely slowly growing function.

**Proposition 8.2.3.** *Let $t : \mathbb{R}^+ \to \mathbb{R}^+$ be a time bound which is continuous, monotone increasing, and unbounded. If a function $f : I \to \mathbb{R}^+$ is bounded by $t$ on $\mu$-average on spheres, then $f$ is generically $t(x \log^* x)$ relative to the asymptotic density $\rho_\mu$.*

*Proof.* Suppose a function $f : I \to \mathbb{R}^+$ is bounded by $t$ on $\mu$-average on spheres, i.e., for some $c$ and every $n$,

$$\int_{I_n} t^{-1}(f(x))\mu_n(x) \le cn.$$

Let $S(f,t) = \{x \in I \mid f(x) > t(s(x) \log^* s(x))\}$. Then

$$\mu_n(S(f,t) \cap I_n) = \mu_n(\{x \in I_n \mid t^{-1}(f(x)) > s(x) \log^* x\}) \le \frac{cn}{n \log^* n} \to \infty,$$

so $S(f,t)$ is a generic subset of $I$ with respect to the asymptotic density $\rho_\mu$. Hence, $t(x \log^* x)$ is a generic upper bound for $f$, as required. $\qquad\square$

### 8.2.3 When generically easy implies easy on average

Generic algorithms provide a powerful tool to study average case complexity of algorithmic problems. In Proposition 8.2.4 below we give a sufficient condition for an algorithmic problem to have a decision algorithm that runs within a given bound on average. The first applications of these methods had appeared in [75].

Recall that a function $h : I \to \mathbb{R}^+$ is $(s,\mu)$-balanced on a subset $K \subseteq I$ if there exists an $s$-upper bound $U : \mathbb{R}^+ \to \mathbb{R}^+$ (so $h(w) \le U(s(w))$), which is $(s,\mu)$ balanced on $K$, i.e.,

$$\Sigma_n U(n)n^{-1}\mu(K \cap I_n) < \infty.$$

**Proposition 8.2.4 (Generic test).** *Suppose a problem $\mathcal{D}$ is decidable in time $f(n)$ by a total decision algorithm $\mathcal{B}$, and generically decidable (with respect to the asymptotic density $\rho_\mu$) in time $f_1(n)$ by a partial algorithm $\mathcal{B}_1$ on a generic set $K \subseteq I$. If the function $f_1^{-1}(f(n))$ is $(s,\mu)$-balanced on $I - K$, then $\mathcal{D}$ is decidable on $\mu$-average in time bounded by $f_1(n)$.*

*Proof.* In the notation above define $\mathcal{A}$ to be the algorithm consisting of running $\mathcal{B}$ and $\mathcal{B}_1$ concurrently. Clearly, $\mathcal{A}$ is a decision algorithm for $\mathcal{D}$. Observe that $T_{\mathcal{A}}(w) \leq f_1(s(w))$ if $w \in K$, and $T_{\mathcal{A}}(w) \leq f(s(w))$ otherwise. By Proposition (7.2.22), the algorithm $\mathcal{A}$ runs within the time bound $f_1$ on $\mu$-average. The result follows.                                                                                              □

Now that we have shown two particular applications of the Generic test, one can construct more examples in a similar fashion.

Recall that a nonnegative function $f(n)$ is *subexponential* if for any $r > 1$ we have
$$\lim_{n \to \infty} \frac{f(n)}{r^n} = 0.$$

Note that this implies that for every $r > 1$,
$$\sum_{n=1}^{\infty} \frac{f(n)}{r^n} < \infty.$$

Also, $f(n)$ is *superpolynomial* if for every $k > 1$,
$$\lim_{n \to \infty} \frac{n^k}{f(n)} = 0.$$

In this case for every $k > 1$,
$$\sum_{n=1}^{\infty} \frac{n^k}{f(n)} < \infty.$$

**Corollary 8.2.5.** *Suppose a problem $\mathcal{D}$ satisfies the conditions of Proposition 8.2.4. Then:*

1) *if $f_1$ is a polynomial, $f_1(x) = cx^m$, $f$ is subexponential, and the residual probability function $\delta_{\bar{K}} : n \to \mu(\bar{K} \cap I_n)$ for $K$ approaches zero exponentially fast, then $\mathcal{D}$ has polynomial on $\mu$-average time complexity.*

2) *if $f_1$ is exponential, $f_1(x) = a^x$, where $a > 1$, $f$ is superexponential, $f(x) = a^{x^k}$ for some $k > 1$, and the residual probability function $\delta_{\bar{K}} : n \to \mu(\bar{K} \cap I_n)$ for $K$ approaches zero superpolynomially fast, then $\mathcal{D}$ has exponential on $\mu$-average time complexity.*

*Proof.* To prove 1) it suffices to show that $f_1^{-1}(T_{\mathcal{A}}(w))$ has an $(s, \mu)$-balanced $s$-upper bound. Put $U = f_1^{-1}(f(w))$, then $f_1^{-1}(T_{\mathcal{A}}(w)) \leq f_1^{-1}(f(s(w))) = U(s(w))$, so $U$ is an $s$-upper bound for $f_1^{-1}(T_{\mathcal{A}}(w))$. Note that $U(n) = (c^{-1}f(w))^{\frac{1}{k}}$ is subexponential. Since the residual probability function $\delta_K : n \to \mu(\bar{K} \cap I_n)$ for $K$ approaches zero exponentially fast, this implies that there exist constants $0 < q < 1$ and $0 < d$ such that for every $n$,
$$\mu(\bar{K} \cap I_n) \leq dq^n.$$

Therefore,
$$\Sigma_n U(n)n^{-1}\mu(\bar{K} \cap I_n) \leq d\Sigma_n U(n)q^n < \infty,$$

so $U$ is $(s,\mu)$-balanced, as claimed.

2) Similarly, if we put $U(n) = n^k$, then

$$f_1^{-1}(T_A(w)) \leq f_1^{-1}(f(s(w)) = \log_a(a^{(s(w)^k)}) = s(w)^k = U(s(w)),$$

so $U$ is an $s$-upper bound for $f_1^{-1}(T_A(w))$. Clearly, $U$ is $(s,\mu)$-balanced, since

$$\Sigma_n U(n)n^{-1}\mu(\bar{K} \cap I_n) \leq \Sigma_n n^{k-1}\mu(\bar{K} \cap I_n) < \infty,$$

as required. □

# Chapter 9

# Generic Complexity of NP-complete Problems

In this chapter, following [48], we show that the **NP**-complete problems Subset Sum and 3-Sat have low generic complexity. As it is well known that these problems are easy most of the time, these results confirm our expectations. The difficult instances are rare; we discuss them, too.

## 9.1 The linear generic time complexity of subset sum problem

There are two versions of the subset sum problem. The input in both is a sequence of natural numbers $c, w_1, \ldots, w_n$. The decision problem is to determine if the equation $\sum x_i w_i = c$ has a solution for $x_i = 0, 1$. The optimization problem is to maximize $s = \sum x_i w_i$ subject to $s \leq c$ and $x_i = 0, 1$.

The subset sum decision problem is **NP**-complete, but the optimization problem is routinely solved in practice and gives a solution to the decision problem. It seems that difficult instances of subset sum are rare. We will show that there is a linear time algorithm which works on a generic set of inputs. For a complete discussion of the subset sum problem see the survey [79].

Our first problem here is to define a generic set. When discussing **NP**-completeness, the natural size of an instance is the number of bits in the input or something close to that. Accordingly, we let $B_n$ be the set of inputs of length $n$.

Let us be more precise now. Pick an alphabet $\{0, 1, \hat{1}\}$ and consider the language $L$ of all strings beginning with $\hat{1}$. For any string in $L$, a substring which begins with $\hat{1}$ and continues up to but not including the next $\hat{1}$ (or up to the end of the string) is interpreted as a binary number by taking $\hat{1} = 1$. In this way each element of $L$ yields a sequence of binary numbers. The first number is $c$ and the subsequent ones are the $w_i$'s. For example, $c = 5$, $w_1 = 3$, $w_2 = 4$ is encoded as $a01a1a00$.

The definition of a generic set is now clear: $B_n$, the ball of radius $n$, consists of all words in $L$ of length at most $n$, and a subset $M \subset L$ is generic if

$$\lim_{n \to \infty} \frac{|M \cap B_n|}{|B_n|} = 1.$$

We will use the following sufficient condition for $M$ to be generic. Let $S_n$ be the sphere of radius $n$; that is, $S$ consists of the $3^{n-1}$ strings in $L$ of length exactly $n$. It is straightforward to show that $M$ is generic if

$$\lim_{n \to \infty} \frac{|M \cap S_n|}{|S_n|} = 1.$$

Our algorithm is trivial. Given input $c, w_1, \ldots, w_n$, it checks if $c$ is equal to some $w_i$. If so, the algorithm outputs "Yes" and otherwise "Don't know". To complete our analysis we need to check that the set of inputs for which $c$ matches some $w_i$ is generic.

We will bound the fraction (or measure) of inputs in $S_n$ for which there is no match. Let $w$ be an input string of length $n$. If $w$ contains no $\hat{1}$'s except the first one, then there is no match because there are no $w_i$'s. The number of $w$'s of this sort is $2^{n-1}$ and their measure in $S_n$ is $(\frac{2}{3})^{n-1}$.

Next consider inputs whose first prefix of the form $\hat{1}(0+1)^*\hat{1}$ is of length at least $m = \lfloor (\log n)/2 \log 3 \rfloor$. The measure of this set is bounded from above by

$$\sum_{k=m}^{\infty} (\frac{2}{3})^k = (\frac{2}{3})^m (1 - \frac{2}{3})^{-1} = 3(\frac{2}{3})^m \leq 3(\frac{2}{3})^{(\log n)/2 \log 3 - 1} \leq \frac{9}{4} n^{-1/5}.$$

Finally suppppose the input $w$ has first prefix of length $k$, $2 \leq k < m$. Divide $k$ into $n$: $n = kq + r$, $0 \leq r < k$. Now $w$ is a product of $q$ subwords of length $k$ followed by a suffix of length $r$. Since the first block is not duplicated, the number of possibilities for each subsequent block is $(3^k - 1)$. Thus the number of $w$'s of this form without a match for $c$ is at most $2^{k-2}(3^k - 1)^{q-1} 3^r \leq (3^k - 1)^q 3^r$. The measure is bounded from above by

$$(\frac{1}{3})^{n-1}(3^k - 1)^q 3^r = 3(1 - (\frac{1}{3})^k)^q \leq 3(1 - (\frac{1}{3})^k)^{(n/k)-1} \leq 3(1 - (\frac{1}{3})^m)^{(n/m)-1}.$$

Now $e^{-x} \geq 1 - x$ for all $x$ implies $(1 - \frac{1}{x})^x \leq 1/e$. Thus

$$(1 - (\frac{1}{3})^m)^{(n/m)-1} \leq (\frac{1}{e})^{((n/m)-1)/3^m}.$$

But $3^m \leq 3^{(\log n)/(2 \log 3)} = \sqrt{n}$. Therefore,

$$((n/m) - 1)/3^m \geq \sqrt{n}/m - 1/\sqrt{n} \geq (2 \log 3)(\sqrt{n}/\log n) - 1/\sqrt{n} \geq n^{1/3}$$

for $n$ large enough.

We conclude that the measure in $S_n$ of the inputs for which $c$ does not match any $w_i$ is at most

$$(\frac{2}{3})^{n-1} + \frac{9}{4}n^{-1/5} + 3(\frac{1}{e})^{n^{1/3}} = O(n^{-1/5}).$$

## 9.2 A practical algorithm for subset sum problem

We present a more practical algorithm which works well with respect to a different stratification of Subset Sum. Our algorithm is adapted from [26].

Suppose that for some positive integer $b$ the weights are chosen uniformly from the interval $[1, b]$ (in $Z$) and $c$ is chosen uniformly from $[1, nb]$.

**Algorithm 9.2.1.** Compute $w_1 + w_2 + \cdots$ until one of the following happens.

1. If $w_1 + w_2 + \cdots + w_j = c$, say "Yes" and halt.

2. If $w_1 + w_2 + \cdots + w_n < c$, say "No" and halt.

3. If $w_1 + w_2 + \cdots + w_{j-1} < c < w_1 + w_2 + \cdots + w_j$, then

    (a) If $w_1 + w_2 + \cdots + w_{j-1} + w_k = c$ for some $k$ with $j < k \le n$, say "yes" and halt.

    (b) Else say "Don't know" and halt.

Clearly Algorithm 9.2.1 is correct and runs in linear time. Let us estimate the probability of a "Don't know" answer. Since $c$ is chosen uniformly from the interval $[1, nb]$, the probability of $w_1 + w_2 + \cdots + w_{j-1} < c < w_1 + w_2 + \cdots + w_j$ is $(w_j - 1)/nb \le 1/n$. Furthermore, the probability that there is no suitable $x_k$ is $((b-1)/b)^{(n-j)}$. Thus the probability of "Don't know" can be calculated as at most

$$\sum_{j=1}^{n}(1/n)((b-1)/b)^{(n-j)} \le b/n.$$

Now consider the set $W$ of all instances of the subset sum problem and let $X$ be the subset on which Algorithm 9.2.1 returns "yes" or "no". An instance is a tuple consisting of a constraint and a finite sequence of weights, $(c, w_1, w_2, \ldots)$. Stratify $W$ by defining $S_n$, the sphere of radius $n$, to be all tuples $(c, w_1, w_2, \ldots, w_n)$ with $w_i \le \sqrt{n}$ and $c \le n^{3/2}$. In other words, $b = \sqrt{n}$. The probability distribution defined above is the equiprobable measure on $S_n$ whence $|X \cup S_n|/|S_n| \ge 1 - b/n = 1 - 1/\sqrt{n}$. Thus $W$ is generic.

## 9.3 3-Satisfiability

3-SAT is a well-known **NP**-complete problem. An instance of 3-SAT is a finite set of clauses each consisting of the disjunction of three variables or negated variables, e.g.,

$$\{(x_1 \vee \neg x_4 \vee x_7), (\neg x_3 \vee x_4 \vee x_6), \ldots\}. \tag{9.1}$$

The problem is to decide whether or not there is a truth assignment which makes all the clauses true. The corresponding optimization problem is to find a truth assignment which makes as many as possible of the clauses true.

3-SAT has been much investigated; overviews are given in [14, Section 8] and [24]. Various ensembles of probability distributions have been proposed. A popular one is to introduce for each $n$ a parameter $m_n$ and consider only instances which are formed by choosing $m_n$ clauses uniformly at random from the $8\binom{n}{3}$ clauses composed of distinct variables. In our language and under the mild restriction that $m_n$ is strictly increasing as a function of $n$, these are the instances of size $n$ and form the sphere of radius $n$.

The motivation for introducing the parameter $m_n$ is first that for fixed $n$ the difficulty of an instance of 3-SAT seems to depend on the density, $c = m_n/n$, and second in practical applications the density is often more or less constant. Extensive experiments with various algorithms have shown that as $c$ increases from zero, the average runtime increases to a maximum and then decreases [24]. The value of the maximum depends on the particular algorithm but is independent of $n$ [22].

If $m_n$ is large enough ($m_n \geq C(n^{3/2})$ for a sufficiently large constant $C$ suffices), then almost all instances are unsatisfiable; and heuristic refutation procedures suffice to show that 3-SAT $\in$ **AvgP**. It does not seem to be known what happens for lower values of $m_n$.

By Lemma 8.1.7 an algorithm whose time function admits an upper bound $f : I \to \mathbb{N}$ which is spherically polynomial on $\mu$-average is generically polynomial. Thus the proof that 3-SAT $\in$ **AvgP** for $m_n$ large enough probably also yields a proof that 3-SAT $\in$ **GenP** for $m_n$ large enough. Below we present a slightly different approach which uses a crude refutation procedure to show 3-SAT $\in$ **Gen$_{str}$P**.

We describe instances of 3-SAT as words in a language and measure size by word length. An instance of 3-SAT like (9.1) may be specified by the indices of the variables together with an indication of which variables are to be negated. In this notation the instance (9.1) becomes

$$1 \vee \overline{1}00v111\#\overline{1}1 \vee 100 \vee 110\# \cdots \tag{9.2}$$

where the indices are written in binary except that the leading 1 is changed to $\overline{1}$ if the variable is to be negated, and $\#$ indicates the end of a clause.

The set of instances, $I$, is then the set of labels of cycles from vertex $\alpha$ to itself in the graph $\Gamma_1$ given in Figure 9.1. In other words $I$ is the language accepted by the finite automaton $\Gamma_1$ with initial vertex $\alpha$ and terminal vertex $\alpha$. The sphere $I_n$ consists of the labels of cycles from vertex $\alpha$ to itself.

$\Gamma_1$ is deterministic in that no vertex has two out-edges with the same label. It follows that distinct paths starting at the same vertex have distinct labels.

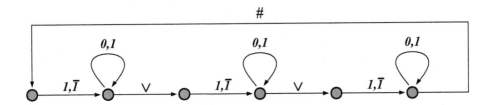

Figure 9.1: A finite automaton $\Gamma$ accepting $I$. An edge with two labels stands for two edges with one label each.

Consequently $|I_n|$, the size of $I_n$, equals the number of cycles of length $n$ through $\alpha$.

There are eight essentially different clauses with variables $x_1, x_2, x_3$. As the variables in a clause may occur in any order, there are strictly speaking forty-eight clauses involving $x_1, x_2, x_3$. If they all appear in a particular input, then that input is not satisfiable. Thus the following partial algorithm is correct.

**Algorithm 9.3.1.** Input an instance of 3-SAT
If all the clauses with variables $1, 10, 11$ occur, say "No"
Else loop forever.

**Theorem 9.3.2.** 3-SAT $\in$ **SGP**.

*Proof.* As the partial algorithm runs in linear time (on its halting set), it suffices to show that the set of inputs which omit a single fixed clause with variables $x_1, x_2, x_3$ is strongly negligible. It will follow that the set of inputs which contain all forty-eight clauses on $x_1, x_2, x_3$ is strongly generic.

Let $A_1$ be the adjacency matrix for $\Gamma_1$ in Figure 9.1. Clearly if the vertices of $\Gamma_1$ are partitioned into two nonempty sets, then there are edges from each set to the other. In other words $A_1$ is indecomposable. By [27, Theorem I] $A_1$ has a positive real eigenvalue $r$ of multiplicity 1 which is at least as large as the absolute value of any other eigenvalue of $A_1$. In addition increasing any entry of $A_1$ increases $r$.

The trace of $A^n$ is equal to the sum over all vertices of the number of cycles of length $n$ from each vertex to itself. Since every vertex is on a cycle through $\alpha$, a straightforward argument shows that the size of $I_n$ satisfies $|I_n| = \Theta(r^n)$, that is, $|I_n|$ and $r^n$ have the same order of growth.

Let $X$ be the set of instances in $I$ which omit the clause $x_1 \vee x_3 \vee \neg x_2$. To complete the proof it suffices to show that $I$ is accepted by a deterministic automaton $\Gamma_2$ such that removing an edge yields an automaton $\Gamma_3$ which accepts $X$, whose adjacency matrix $A_3$ is indecomposable, and to which [27, Theorem I] applies. Indeed suppose this is so, and let $A_2$ be the adjacency matrix of $\Gamma_2$. As $A_2$ is obtained by increasing an entry of $A_3$, $A_2$ is also indecomposable. In addition since $\Gamma_2$ is a deterministic automaton accepting $I$, its maximum real eigenvalue is $r$. By [27, Theorem I] the maximum real eigenvalue of $A_3$ is $s < r$. Thus $|X \cap I_n|/|I_n| = O((s/r)^n)$. Hence $X$ is negligible as required.

We leave it to the reader to check that we may take $\Gamma_2$ to be the automaton in Figure 9.2. $\Gamma_3$ is obtained by removing the outedge with label $\#$ at vertex $\beta$. $\square$

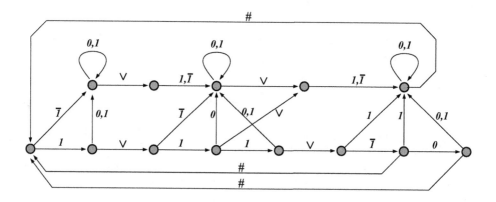

Figure 9.2: Another finite automaton $\Gamma$ accepting $I$. Vertex $\alpha$ is the initial vertex and the single terminal vertex.

Most modern cryptographic schemes use algebraic systems such as rings, groups, lattices, etc. as platforms. Typically, cryptographic protocols involve a random choice of various algebraic objects related to the platforms, like elements, or subgroups, or homomorphisms. One of the key points in using randomness is to foil various statistical attacks, or attacks which could use some specific properties of objects if the latter are not chosen randomly. The main goal of this part of the book is to show that randomly chosen objects quite often have very special properties, which yields some "unexpected" attacks. We argue that the knowledge of basic properties of random objects must be part of any serious cryptanalysis and it has to be taken into consideration while choosing "good" keys.

In the paper [105] the authors study asymptotic properties of words representing the trivial element in a given finitely presented group $G$. It turns out that a randomly chosen trivial word in $G$ has a "hyperbolic" van Kampen diagram, even if the group $G$ itself is not hyperbolic. This allows one to design a correct (no errors) search/decision algorithm which solves the word (search) problem in polynomial time on a generic subset (i.e., on "most" elements) of a given group $G$. A similar result for the conjugacy search problem in finitely presented groups has been proven in [104] (we refer to [138] for a detailed analysis of the generic complexity of the word and conjugacy search problems in groups). These results show that group-based cryptographic schemes whose security is based on the word or conjugacy search problems are subject to effective attacks, unless the keys are chosen in the complements of the corresponding generic sets.

In this part of the book we study asymptotic properties of finitely generated subgroups of groups. We start by introducing a methodology to deal with asymptotic properties of subgroups in a given finitely generated group, then we describe two such properties, and finally we show how one can use them in cryptanalysis of group-based cryptographic schemes. Then we dwell on the role of asymptotically dominant properties of subgroups in modern cryptanalysis. We mostly focus on one particular example, the Anshel-Anshel-Goldfeld (AAG) key establishment protocol [1]; however, it seems plausible that similar analysis can be applied to some other cryptographic schemes as well. One of our main goals here is to give mathematical reasons behind surprisingly high success rates of the so-called *length based attacks* in breaking the AAG protocol. Another goal is to analyze an attack that we call a *quotient attack* (cf. our Section 6.1.6). We also want to emphasize that we believe that this "asymptotic cryptanalysis" suggests a way to choose strong keys (like groups, subgroups, or elements) for the general AAG scheme (with different groups as the platforms) that may foil some of the known attacks, including the ones discussed here.

Our main focus in this part of the book is on the length-based attacks (LBA). This idea appeared first in the paper [67] by Hughes and Tannenbaum, and later was further developed in two papers [42] and [43] by Garber, Kaplan, Teicher, Tsaban, and Vishne. Recently, the most successful version of this attack for the (simultaneous) conjugacy search problem in braid groups was developed in [112]. We note that Ruinsky, Shamir, and Tsaban used LBA in attacking some other algorithmic problems in groups [118]. Yet more recently, LBA was used in [111] and [73] to attack a "lightweight" cryptographic protocol due to Anshel-Anshel-Goldfeld-Lemieux [2].

The basic idea of LBA is very simple. One solves the simultaneous conjugacy search problem relative to a subgroup (SCSP*) (i.e., with a constraint that solutions should be in a given subgroup) precisely the same way as this would be done in a free group. Surprisingly, experiments show that this strategy works well in groups which seem to be far from being free, for instance, in braid groups. We claim that the primary reason behind this phenomenon is that asymptotically, finitely generated subgroups in many groups are free. More precisely, in many groups a randomly chosen finite set of elements freely generates a free subgroup with overwhelming probability. We say that such groups have the *free basis property*. This allows one to analyze the generic case complexity of LBA, SCSP*, and some other related algorithmic problems. Moreover, we argue that LBA implicitly relies on fast computing of the geodesic length of elements in finitely generated subgroups of the platform group, or of some good approximations of that length. In fact, most LBA strategies tacitly assume that the geodesic length of elements in a given group is a good approximation of the geodesic length of the same elements in a subgroup. At first it may look like a wrong assumption, because it is known that even in a braid group $B_n$, $n \geq 3$, there are infinitely many subgroups whose distortion function (which measures the geodesic length of elements in a subgroup relative to that in $G$) is not bounded by any recursive function. We show, nevertheless, that in many groups the distortion of randomly chosen finitely generated subgroups is linear. Our main focus is on the braid group $B_n$, $n \geq 3$. We establish our main results here for the pure braid group $PB_n$, which is a subgroup of finite index in the ambient braid group $B_n$. We conjecture that the results hold in the group $B_n$ as well, and hope to fill in this gap in the near future. We also mention that our results hold for all finitely generated groups that have nonabelian free quotients.

While studying length based attacks, we realized that quotient attacks (QA) appear to be yet another class of powerful attacks on the AAG scheme. These attacks are just fast generic algorithms to solve various search problems in groups, cf. our Section 6.1.6. The main idea behind QA is the following observation: to solve a computational problem in a group $G$ it suffices, on most inputs, to solve it in a suitable quotient $G/N$, provided $G/N$ has a fast decision algorithm for the problem. Robustness of such an algorithm relies on the following property of the quotient $G/N$: a randomly chosen finitely generated subgroup of $G$ has trivial intersection with the kernel $N$. In particular, this is the case if $G/N$ is a free nonabelian group. We note that a similar idea was already exploited in [75], but there the answer was given only for inputs in the "No" part of a given decision problem, which does not apply to search problems at all. The strength of our approach comes from the extra requirement that $G/N$ has the free basis property.

To conclude these introductory remarks, we say that we believe that the AAG scheme, despite being heavily battered by several attacks, is still very much "alive". It simply did not get a fair chance to survive because of insufficient group-theoretic research it required. It is still quite plausible that there are platform groups and methods to choose strong keys for AAG that would foil all known attacks. To find such a platform group is an interesting and challenging algebraic problem. We emphasize once again that our method of analyzing generic case complexity of computational security assumptions of the AAG scheme, based on the asymptotic behavior of subgroups in a given group, creates a bridge between asymptotic algebra and cryptanalysis. Also, this method can be applied to some other cryptographic schemes as well.

# Chapter 10

# Asymptotically Dominant Properties

In this chapter we develop some tools to study asymptotic properties of subgroups of groups. Throughout this chapter, by $G$ we denote a group with a finite generating set $X$.

## 10.1 A brief description

Asymptotic properties of subgroups, a priori, depend on a given probability distribution on these subgroups. In general, there are several natural procedures to generate subgroups in a given group. However, there is no unique universal distribution of this kind. We refer to [101] for a discussion on different approaches to random subgroup generation.

Our basic principle here is that in applications, one has to consider a particular distribution that comes from a particular problem addressed in the given application, say in a cryptographic protocol. As soon as the distribution is fixed, one can approach asymptotic properties of subgroups via asymptotic densities with respect to a fixed stratification of the set of subgroups. We briefly discuss these ideas below and refer to [5, 17, 75, 76], and to a recent survey [48], for a thorough exposition. In Section 10.2, we adjust these general ideas to a particular way to generate subgroups that is used in cryptography.

The first step is to choose and fix a particular way to describe finitely generated subgroups $H$ of the group $G$. For example, a description $\delta$ of $H$ could be a tuple of words $(u_1, \ldots, u_k)$ in the alphabet $X^{\pm 1} = X \cup X^{-1}$ representing a set of generators of $H$, or a folded finite graph that accepts the subgroup generated by the generators $\{u_1, \ldots, u_k\}$ of $H$ in the ambient free group $F(X)$ (see [74]), etc. In general, the descriptions above are by no means unique for a given subgroup $H$.

When the way of describing subgroups in $G$ is fixed, one can consider the set $\Delta$ of all such descriptions of all finitely generated subgroups of $G$. The next step is to define a *size* $s(\delta)$ of a given description $\delta \in \Delta$, i.e., a function

$$s : \Delta \to \mathbb{N}$$

in such a way that the set (the ball of radius $n$)

$$B_n = \{\delta \in \Delta \mid s(\delta) \le n\}$$

is finite. This gives a *stratification* of the set $\Delta$ into a union of finite balls:

$$\Delta = \cup_{n=1}^{\infty} B_n. \qquad (10.1)$$

Let $\mu_n$ be a given probabilistic measure on $B_n$ (it could be the measure induced on $B_n$ by some fixed measure on the whole set $\Delta$ or a measure not related to any measure on $\Delta$). The stratification (10.1) and the collection of measures

$$\{\mu_n\} = \{\mu_n \mid n \in \mathbb{N}\} \qquad (10.2)$$

allow one to estimate the asymptotic behavior of subsets of $\Delta$. For a subset $R \subseteq \Delta$, the *asymptotic density* $\rho_\mu(R)$ is defined by the following limit (if it exists),

$$\rho_\mu(R) = \lim_{n \to \infty} \mu_n(R \cap B_n).$$

If $\mu_n$ is the uniform distribution on the finite set $B_n$, then

$$\mu_n(R \cap B_n) = \frac{|R \cap B_n|}{|B_n|}$$

is the $n$-th *frequency*, or probability, to hit an element from $R$ in the ball $B_n$. In this case we refer to $\rho_\mu(R)$ as to the *asymptotic density* of $R$ and denote it by $\rho(R)$.

One can also define the asymptotic density using $\limsup$ rather than $\lim$, in which case $\rho_\mu(R)$ does always exist.

We say that a subset $R \subseteq \Delta$ is *generic* if $\rho_\mu(R) = 1$ and *negligible* if $\rho_\mu(R) = 0$. It is worthwhile to mention that the asymptotic densities not only allow one to distinguish between "large" (generic) and "small" (negligible) sets, but also give a tool to distinguish various large (or small) sets. For instance, we say that $R$ has asymptotic density $\rho_\mu(R)$ with a *super-polynomial convergence rate* if

$$|\rho_\mu(R) - \mu_n(R \cap B_n)| = o(n^{-k})$$

for any $k \in \mathbb{N}$. For brevity, $R$ is called *strongly generic* if $\rho_\mu(R) = 1$ with a super-polynomial convergence rate. The set $R$ is *strongly negligible* if its complement $S - R$ is strongly generic.

Similarly, one can define exponential convergence rates and exponentially generic (negligible) sets.

## 10.2 Random subgroups and generating tuples

In this section we review a procedure, which is most commonly used in group-based cryptography to generate random subgroups of a given group (see e.g., [1]). Briefly, the procedure is as follows.

**Random generation of subgroups in $G$:**

- pick a random $k \in \mathbb{N}$ within given bounds: $K_0 \le k \le K_1$;

- pick randomly $k$ words $w_1, \ldots, w_k \in F(X)$ with fixed length range: $L_0 \le |w_i| \le L_1$;

- output the subgroup $\langle w_1, \ldots, w_k \rangle$ of $G$.

Without loss of generality we may fix, from the beginning, a single natural number $k$, instead of choosing it from a finite interval $[K_0, K_1]$ (by the formula of complete probability the general case can be reduced to this one). Fix $k \in \mathbb{N}$, $k \ge 1$, and focus on the set of all $k$-generated subgroups of $G$.

The corresponding descriptions $\delta$, the size function, and the corresponding stratification of the set of all descriptions can be formalized as follows. By a description $\delta(H)$ of a $k$-generated subgroup $H$ of $G$ we understand here any $k$-tuple $(w_1, \ldots, w_k)$ of words from $F(X)$ that generates $H$ in $G$. Hence, in this case the space of all descriptions is the Cartesian product $F(X)^k$ of $k$ copies of $F(X)$:

$$\Delta = \Delta_k = F(X)^k.$$

The size $s(w_1, \ldots, w_k)$ can be defined either as the total length of the generators

$$s(w_1, \ldots, w_k) = |w_1| + \ldots + |w_k|,$$

or as the maximum length of the components:

$$s(w_1, \ldots, w_k) = \max\{|w_1|, \ldots, |w_k|\}.$$

Our approach works for both definitions, so we do not specify which one we use here. For $n \in \mathbb{N}$, denote by $B_n$ the ball of radius $n$ in $\Delta$:

$$B_n = \{(w_1, \ldots, w_k) \in F(X)^k \mid s(w_1, \ldots, w_k) \le n\}.$$

This gives the required stratification

$$\Delta = \cup_{n=1}^{\infty} B_n.$$

For a subset $M$ of $\Delta$ we define the asymptotic density $\rho(M)$ relative to the stratification above assuming the uniform distribution on the balls $B_n$:

$$\rho(M) = \lim_{n \to \infty} \frac{|B_n \cap M|}{|B_n|}.$$

We note that there are several obvious deficiencies in this approach: (1) we consider subgroups with a fixed number of generators; (2) a subgroup typically has many different $k$-generating tuples; (3) every generator can be described by several different words from $F(X)$. Thus, our descriptions are far from being unique. However, as we have mentioned before, these models reflect standard methods to generate subgroups in cryptographic protocols. We refer to [101] for other approaches.

## 10.3   Asymptotic properties of subgroups

Let $G$ be a group with a finite set of generators $X$, and $k$ a fixed positive integer. Denote by $\mathcal{P}$ a property of descriptions of $k$-generated subgroups of $G$. By $\mathcal{P}(G)$ we denote the set of all descriptions from $\Delta = \Delta_k$ that satisfy $\mathcal{P}$ in $G$.

**Definition 10.3.1.** We say that a property $\mathcal{P} \subseteq \Delta$ of descriptions of $k$-generated subgroups of $G$ is:

1)  *asymptotically visible* in $G$ if $\rho(\mathcal{P}(G)) > 0$;

2)  *generic* in $G$ if $\rho(\mathcal{P}(G)) = 1$;

3)  *strongly generic* in $G$ if $\rho(\mathcal{P}(G)) = 1$, and the rate of convergence of $\rho_n(\mathcal{P}(G))$ is super-polynomial;

4)  *exponentially generic* in $G$ if $\rho(\mathcal{P}(G)) = 1$ and the rate of convergence of $\rho_n(\mathcal{P}(G))$ is exponential.

Informally, if $\mathcal{P}$ is asymptotically visible for $k$-generated subgroups of $G$, then there is a certain nonzero probability that a randomly and uniformly chosen description $\delta \in \Delta$ of a sufficiently big size results in a subgroup of $G$ satisfying $\mathcal{P}$. Similarly, if $\mathcal{P}$ is exponentially generic for $k$-generated subgroups of $G$, then a randomly and uniformly chosen description $\delta \in \Delta$ of a sufficiently big size results in a subgroup of $G$ satisfying $\mathcal{P}$ with overwhelming probability. Likewise, one can interpret generic and strongly generic properties of subgroups. If a set of descriptions $\Delta$ of subgroups of $G$ is fixed, then we sometimes abuse the terminology and refer to asymptotic properties of descriptions of subgroups as asymptotic properties of the subgroups itself.

**Example 10.3.2.** Let $H$ be a fixed $k$-generated group. Consider the following property $\mathcal{P}_H$: a given description $(w_1, \ldots, w_k) \in F(X)^k$ satisfies $\mathcal{P}_H$ if the subgroup $\langle w_1, \ldots, w_k \rangle$ generated in $G$ by this tuple, is isomorphic to $H$. If $\mathcal{P}_H(G)$ is asymptotically visible (generic) in $\Delta$, then we say that the group $H$ is asymptotically visible (generic) in $G$ (among $k$-generated subgroups).

By $k$-*spectrum* $Spec_k(G)$ of $G$ we denote the set of all (up to isomorphism) $k$-generated subgroups that are asymptotically visible in $G$.

There are several natural problems about asymptotically visible subgroups of $G$ that play an important role in cryptography. For example, when choosing

$k$-generated subgroups of $G$ randomly, it might be useful to know what kind of subgroups you can get with nonzero probability. Hence the following problem is of interest:

**Problem 10.3.3.** What is the spectrum $Spec_k(G)$ for a given group $G$ and a natural number $k \geq 1$?

More technical, but also important in applications, is the following problem.

**Problem 10.3.4.** Does the spectrum $Spec_k(G)$ depend on a given finite set of generators of $G$?

We will see in due course that knowing the answers to these problems is important for choosing strong keys in some group-based cryptographic schemes.

## 10.4 Groups with generic free basis property

**Definition 10.4.1.** We say that a tuple $(u_1, \ldots, u_k) \in F(X)^k$ has the *free basis* property ($\mathcal{FB}$) in $G$ if it freely generates a free subgroup in $G$.

In [71] Jitsukawa showed that $\mathcal{FB}$ is generic for $k$-generated subgroups of a finitely generated free nonabelian group $F(X)$ for every $k \geq 1$ with respect to the standard basis $X$. Martino, Turner and Ventura strengthened this result in [98] by proving that $\mathcal{FB}$ is *exponentially generic* in $F(X)$ for every $k \geq 1$ with respect to the standard basis $X$. Recently, it has been shown in [101] that $\mathcal{FB}$ is exponentially generic in arbitrary hyperbolic nonelementary (in particular, free nonabelian) groups for every $k \geq 1$ and with respect to any finite set of generators.

We say that a group $G$ has the *generic free basis* property if $\mathcal{FB}$ is generic in $G$ for every $k \geq 1$ and every finite generating set of $G$. In a similar way, we define groups with *strongly* and *exponentially* generic free basis property. By $\mathcal{FB}_{gen}$, $\mathcal{FB}_{st}$, $\mathcal{FB}_{exp}$, we denote classes of finitely generated groups with, respectively, generic, strongly generic, and exponentially generic free basis property.

The following result gives a host of examples of groups with generic $\mathcal{FB}$.

**Theorem 10.4.2.** *Let $G$ be a finitely generated group, and $N$ a normal subgroup of $G$. If the quotient group $G/N$ is in the class $\mathcal{FB}_{gen}$, or in $\mathcal{FB}_{st}$, or in $\mathcal{FB}_{exp}$, then the whole group $G$ is in the same class.*

*Proof.* Let $H = G/N$, and let $\varphi : G \to H$ be the canonical epimorphism. Fix a finite generating set $X$ of $G$ and a natural number $k \geq 1$. Clearly, $X^\varphi$ is a finite generating set of $H$. By our assumption, the free basis property is generic in $H$ with respect to the generating set $X^\varphi$ and given $k$. By identifying $x \in X$ with $x^\varphi \in H$ we may assume that a finitely generated subgroup $A$ of $G$ and the subgroup $A^\varphi$ have the same set of descriptions. Observe now that a subgroup $A$ of $G$ generated by a $k$-tuple $(u_1, \ldots, u_k) \in F(X)^k$ has the following property: if $A^\varphi$

is free with basis $(u_1^\varphi, \ldots, u_k^\varphi)$, then $A$ is free with basis $(u_1, \ldots, u_k)$. Therefore, for each $t \in \mathbb{N}$ one has

$$\frac{|B_t \cap \mathcal{FB}(G)|}{|B_t|} \geq \frac{|B_t \cap \mathcal{FB}(H)|}{|B_t|}.$$

This implies that if $\mathcal{FB}(H)$ is generic in $H = G/N$, then $\mathcal{FB}(G)$ is generic in $G$, and its convergence rate in $G$ is not less than the corresponding convergence rate in $H$, as claimed.                                                          $\square$

The result above bears on some infinite groups recently used in group-based cryptography. Braid groups $B_n$ appear as one of the main platforms in braid-group cryptography so far (see [1, 84, 32, 2]). Recall (see our Section 5.1) that the braid group $B_n$ can be defined by the classical Artin presentation:

$$B_n = \left\langle\ \sigma_1, \ldots, \sigma_{n-1}\ \middle|\ \begin{array}{ll} \sigma_i\sigma_j\sigma_i = \sigma_j\sigma_i\sigma_j & \text{if } |i-j| = 1 \\ \sigma_i\sigma_j = \sigma_j\sigma_i & \text{if } |i-j| > 1 \end{array}\ \right\rangle.$$

Denote by $\sigma_{i,i+1}$ the transposition $(i, i+1)$ in the symmetric group $\Sigma_n$. The map $\sigma_i \to \sigma_{i,i+1}$, $i = 1, \ldots, n$ gives rise to the canonical epimorphism $\pi : B_n \to \Sigma_n$. The kernel of $\pi$ is a subgroup of index $n!$ in $B_n$, called the *pure braid* group $PB_n$.

**Corollary 10.4.3.** *The free basis property is exponentially generic in the pure braid group $PB_n$ for any $n \geq 3$.*

*Proof.* It is known (see e.g. [10]) that the group $PB_n$, $n \geq 3$, maps onto $PB_3$, and the group $PB_3$ is isomorphic to $F_2 \times \mathbb{Z}$. Therefore, $PB_n$, $n \geq 3$, maps onto the free group $F_2$. Now the result follows from Theorem 10.4.2 and the strong version of Jitsukawa's result mentioned above [98, 101, 71].                        $\square$

Although we have shown that the pure braid group $PB_n$, $n \geq 3$, has exponentially generic free basis property and is a subgroup of finite index in the braid group $B_n$, at the moment we cannot prove that $B_n$ has exponentially generic free basis property, so we ask:

**Problem 10.4.4.** Is it true that the braid group $B_n$, $n \geq 3$, has exponentially generic free basis property?

In [97], partially commutative groups were proposed as possible platforms for some cryptographic schemes. We refer to [9] for more recent discussion on this. By definition, a partially commutative group $G(\Gamma)$ (also known as right angled Artin group, cf. our Section 5.6) is a group associated with a finite graph $\Gamma = (V, E)$, with a set of vertices $V = \{v_1, \ldots, v_n\}$ and a set of edges $E \subseteq V \times V$, given by the following presentation:

$$G(\Gamma) = \langle v_1, \ldots, v_n \mid v_iv_j = v_jv_i \text{ for } (v_i, v_j) \in E \rangle.$$

Note that the group $G(\Gamma)$ is abelian if and only if the graph $\Gamma$ is complete.

**Corollary 10.4.5.** *The free basis property is exponentially generic in nonabelian partially commutative groups.*

*Proof.* Let $G = G(\Gamma)$ be a nonabelian partially commutative group corresponding to a finite graph $\Gamma$. Then there are three vertices in $\Gamma$, say $v_1, v_2, v_3$ such that the subgraph $\Gamma_0$ of $\Gamma$ on these vertices is not a triangle. In particular, the partially commutative group $G_0 = G(\Gamma_0)$ is either a free group $F_3$ (no edges in $\Gamma_0$), or $(\mathbb{Z} \times \mathbb{Z}) * \mathbb{Z}$ (only one edge in $\Gamma_0$), or $F_2 \times \mathbb{Z}$ (precisely two edges in $\Gamma_0$). In all three cases the group $G(\Gamma_0)$ maps onto $F_2$. Now it suffices to observe that $G(\Gamma)$ maps onto $G(\Gamma_0)$ since $G(\Gamma_0)$ is obtained from $G(\Gamma)$ by adding to the standard presentation of $G(\Gamma)$ all the relations of type $v = 1$, where $v$ is a vertex of $\Gamma$ different from $v_1, v_2, v_3$. This shows that $F_2$ is a quotient of $G(\Gamma)$ and the result follows from Theorem 10.4.2. $\qquad\square$

Note that some other groups that have been proposed as platforms in group-based cryptography do not have nonabelian free subgroups at all, so they do not have free basis property for $k \geq 2$. For instance, in [117] Grigorchuk's groups (see our Section 5.7) were used as platforms. Since these groups are periodic (i.e., every element has finite order), they do not contain free subgroups. It is not clear what the asymptotically visible subgroups in Grigorchuk groups are. As another example, note that in [122], the authors propose Thompson's group $F$ as a platform. It is known that there are no nonabelian free subgroups in $F$ (see, for example, [19]), so $F$ does not have free basis property. Recently, some interesting results on the spectrum $Spec_k(F)$ were obtained in [21].

## 10.5 Quasi-isometrically embedded subgroups

In this section we discuss another property of subgroups of $G$ that plays an important role in our cryptanalysis of group-based cryptographic schemes.

Let $G$ be a group with a finite generating set $X$. The *Cayley graph* $\Gamma(G, X)$ is an $X$-labeled directed graph with the vertex set $G$ and such that any two vertices $g, h \in G$ are connected by an edge from $g$ to $h$ with a label $x \in X$ if and only if $gx = h$ in $G$. For convenience we usually assume that the set $X$ is closed under inversion, i.e., $x^{-1} \in X$ for every $x \in X$. One can introduce a metric $d_X$ on $G$ setting $d_X(g, h)$ equal to the length of a shortest word in $X = X \cup X^{-1}$ representing the element $g^{-1}h$ in $G$. It is easy to see that $d_X(g, h)$ is equal to the length of a shortest path from $g$ to $h$ in the Cayley graph $\Gamma(G, X)$. This turns $G$ into a metric space $(G, d_X)$. By $l_X(g)$ we denote the length of a shortest word in generators $X$ representing the element $g$. Clearly, $l_X(g) = d_X(1, g)$.

Let $H$ be a subgroup of $G$ generated by a finite set $Y$ of elements. Then there are two metrics on $H$: one is $d_Y$ described above and the other one is the metric $d_X$ induced from the metric space $(G, d_X)$ on the subspace $H$. The following notion allows one to compare these metrics. Recall that a map $f : M_1 \to M_2$ between two metric spaces $(M_1, d_1)$ and $(M_2, d_2)$ is a *quasi-isometric embedding* if there are constants $\lambda > 1$, $c > 0$ such that for all elements $x, y \in M_1$ the following inequalities hold:

$$\frac{1}{\lambda}d_1(x, y) - c \leq d_2(f(x), f(y)) \leq \lambda d_1(x, y) + c. \tag{10.3}$$

In particular, we say that a subgroup $H$ with a finite set of generators $Y$ is *quasi-isometrically embedded* into $G$ if the inclusion map $i : H \hookrightarrow G$ is a quasi-isometric embedding $i : (H, d_Y) \to (G, d_X)$. Note that in this case the right-hand inequality in (10.3) always holds, since for all $f, h \in H$ one has

$$d_X(i(f), i(h)) \leq \max_{y \in Y}\{l_X(y)\} \cdot d_Y(f, h).$$

Therefore, the definition of a quasi-isometrically embedded subgroup takes the following simple form (in the notation above).

**Definition 10.5.1.** Let $G$ be a group with a finite generating set $X$, and $H$ a subgroup of $G$ generated by a finite set of elements $Y$. Then $H$ is *quasi-isometrically embedded* in $G$ if there are constants $\lambda > 1$, $c > 0$ such that for all elements $f, h \in H$ the following inequality holds:

$$\frac{1}{\lambda}d_Y(f, h) - c \leq d_X(f, h). \tag{10.4}$$

It follows immediately from the definition that if $X$ and $X'$ are two finite generating sets of $G$, then the metric spaces $(G, d_X)$ and $(G, d_{X'})$ are quasi-isometrically embedded into each other. This implies that the notion of quasi-isometrically embedded subgroups is independent of the choice of finite generating sets in $H$ or in $G$ (though the constants $\lambda$ and $c$ could be different).

**Definition 10.5.2.** Let $G$ be a group with a finite generating set $X$. We say that a tuple $(u_1, \ldots, u_k) \in F(X)^k$ has a $\mathcal{QI}$ (quasi-isometric embedding) property in $G$ if the subgroup it generates in $G$ is quasi-isometrically embedded in $G$.

Denote by $\mathcal{QI}(G)$ the set of all tuples in $F(X)^k$ that satisfy the $\mathcal{QI}$ property in $G$. We say that the property $\mathcal{QI}$ is *generic* in $G$ if $\mathcal{QI}(G)$ is generic in $G$ for every $k \geq 1$ and every finite generating set of $G$. Similarly, we define groups with *strongly* and *exponentially* generic quasi-isometric embedding subgroup property. Denote by $\mathcal{QI}_{gen}$, $\mathcal{QI}_{st}$, $\mathcal{QI}_{exp}$ classes of finitely generated groups with, respectively, generic, strongly generic, and exponentially generic quasi-isometric embedding subgroup property.

It is not hard to see that *every* finitely generated subgroup of a finitely generated free group $F$ is quasi-isometrically embedded in $F$, so $F \in \mathcal{QI}_{exp}$.

The following result gives further examples of groups with quasi-isometric embedding subgroup property.

Let $G \in \mathcal{FB}_{gen} \cap \mathcal{QI}_{gen}$. Note that the intersection of two generic sets $\mathcal{FB}(G) \subseteq F(X)^k$ and $\mathcal{QI}(G) \subseteq F(X)^k$ is again a generic set in $F(X)^k$, so the set

$\mathcal{FB}(G) \cap \mathcal{QI}(G)$ of all tuples $(u_1, \ldots, u_k) \in F(X)^k$ that freely generate a quasi-isometrically embedded subgroup of $G$, is generic in $F(X)^k$. Observe that by the remark above and by the result on free basis property in free groups, $\mathcal{FB}_{gen} \cap \mathcal{QI}_{gen}$ contains all free groups of finite rank. The argument applies also to the strongly generic and exponentially generic versions of the properties. To unify references we will use the following notation: $\mathcal{FB}_* \cap \mathcal{QI}_*$, for $* \in \{gen, st, exp\}$.

**Theorem 10.5.3.** *Let $G$ be a finitely generated group with a quotient $G/N$. If $G/N \in \mathcal{FB}_* \cap \mathcal{QI}_*$, then $G \in \mathcal{FB}_* \cap \mathcal{QI}_*$ for any $* \in \{gen, st, exp\}$.*

*Proof.* Let $G$ be a finitely generated group generated by $X$, $N$ a normal subgroup of $G$ such that the quotient $G/N$ is in $\mathcal{FB}_* \cap \mathcal{QI}_*$. Let $\varphi : G \to G/N$ be the canonical epimorphism. By Theorem 10.4.2, $G \in \mathcal{FB}_*$, so it suffices to show now that $G \in \mathcal{QI}_*$.

Let $H$ be a $k$-generated subgroup with a set of generators $Y = (u_1, \ldots, u_k) \in F(X)^k$. Suppose that $Y \in \mathcal{FB}_*(G/N) \cap \mathcal{QI}_*(G/N)$, i.e., the image $Y^\varphi$ of $Y$ in $G/N$ freely generates a free group quasi-isometrically embedded into $G/N$. Observe first that for every element $g \in G$, one has $l_X(g) \geq l_{X^\varphi}(g^\varphi)$, where $l_{X^\varphi}$ is the length function on $G/N$ relative to the set $X^\varphi$ of generators. Since the subgroup $H^\varphi$ is quasi-isometrically embedded into $G/N$, the metric space $(H^\varphi, d_{Y^\varphi})$ quasi-isometrically embeds into $(G^\varphi, d_{X^\varphi})$. On the other hand, $\varphi$ maps the subgroup $H$ onto the subgroup $H^\varphi$ isomorphically (since both are free groups with the corresponding bases), such that for any $h \in H$, one has $d_Y(h) = d_{Y^\varphi}(h^\varphi)$. Now we can deduce the following inequalities for $g, h \in H$:

$$\frac{1}{\lambda} d_Y(g, h) - c = \frac{1}{\lambda} d_{Y^\varphi}(g^\varphi, h^\varphi) - c \leq d_{X^\varphi}(g^\varphi, h^\varphi) \leq d_X(g, h),$$

where $\lambda$ and $c$ come from the quasi-isometric embedding of $H^\varphi$ into $G/N$. This shows that $H$ is quasi-isometrically embedded in $G$, as required. $\square$

**Corollary 10.5.4.** *The following groups are in $\mathcal{FB}_{exp} \cap \mathcal{QI}_{exp}$:*

1) *Pure braid groups $PB_n$, $n \geq 3$;*

2) *Nonabelian partially commutative groups $G(\Gamma)$.*

*Proof.* The arguments in Corollaries 10.4.3, 10.4.5 show that the groups $PB_n$, $n \geq 3$, and $G(\Gamma)$ are non-commutative and have quotient isomorphic to the free group $F_2$. Now the result follows from Theorems 10.4.2 and 10.5.3. $\square$

# Chapter 11

# Length-Based and Quotient Attacks

Our exposition in this chapter essentially follows [106].

## 11.1 Anshel-Anshel-Goldfeld scheme

In this section we discuss the Anshel-Anshel-Goldfeld (AAG) public key exchange protocol [1] and touch briefly on its security.

### 11.1.1 Description of the Anshel-Anshel-Goldfeld scheme

We have already described the AAG protocol in our Section 4.5, but we briefly describe it here again for the reader's convenience.

Let $G$ be a group with a finite generating set $X$. $G$ is called the *platform* of the scheme. We assume that elements $w$ in $G$ have unique normal forms $\bar{w}$ such that it is hard to recover $w$ from $\bar{w}$, and there is a fast algorithm to compute $\bar{w}$ for any given $w$.

The Anshel-Anshel-Goldfeld key exchange protocol is the following sequence of steps. Alice [Bob, respectively] chooses a random subgroup of $G$,

$$A = \langle a_1, \ldots, a_m \rangle \quad [B = \langle b_1, \ldots, b_n \rangle \text{ respectively}]$$

by randomly choosing generators $a_1, \ldots, a_m$ [$b_1, \ldots, b_n$ respectively] as words in $X^{\pm 1}$, and makes it public. Then Alice [Bob, respectively] chooses randomly a secret element $a = u(a_1, \ldots, a_m) \in A$ [$b = v(b_1, \ldots, b_n) \in B$ respectively] as a product of the generators of $A$ [$B$ respectively] and their inverses, takes the conjugates $b_1^a, \ldots, b_n^a$ [$a_1^b, \ldots, a_m^b$ respectively], diffuses them by taking their normal forms $\overline{b_i^a}$ [$\overline{a_j^b}$ respectively], and makes these normal forms public:

$$\overline{b_1^a}, \ldots, \overline{b_n^a} \quad [\overline{a_1^b}, \ldots, \overline{a_m^b} \text{ respectively}].$$

Afterwards, Alice [Bob, respectively] computes $a^{-1}a^b$ $[(b^a)^{-1}b$ respectively] and takes its normal form. Since

$$a^{-1}a^b = [a, b] = (b^a)^{-1}b,$$

the obtained normal form is the shared secret key.

## 11.1.2   Security assumptions of the AAG scheme

In this section we briefly discuss computational security features of the AAG scheme. Unfortunately, in the original description, the authors of the scheme did not state precisely the security assumptions that should make the scheme difficult to break. Here we dwell on several possible assumptions of this type, that often occur, though sometimes implicitly, in the literature on the AAG scheme.

It appears that the security of the AAG scheme relies on the computational hardness of the following computational problem in group theory:

**AAG Problem:** given the whole public information from the AAG scheme, i.e., the group $G$, the elements $a_1, \ldots, a_m$, $b_1, \ldots, b_n$, and $\overline{b_1^a}, \ldots, \overline{b_n^a}, \overline{a_1^b}, \ldots, \overline{a_n^b}$ in the group $G$, find the shared secret key $[a, b]$.

This problem is not a standard group-theoretic problem, so not much is known about its complexity, and it is quite technical to formulate. So it would be convenient to reduce this problem to some standard algorithmic problem about groups or to a combination of such problems. The following problems seem to be relevant here and they attracted quite a lot of attention recently, especially in the braid groups context (braid groups were the original platform for AAG [1]. We refer to the papers [18], [11], [13], [49], [90], [92]. Nevertheless, the precise relationship between these problems and AAG is unclear, see [123] for more details.

**The Conjugacy Search Problem (CSP):** given $u, v \in G$ such that the equation $u^x = v$ has a solution in $G$, find a solution.

**The Simultaneous Conjugacy Search Problem (SCSP):** given $u_i, v_i \in G$, such that the system $u_i^x = v_i$, $i = 1, \ldots, n$ has a solution in $G$, find a solution.

**The Simultaneous Conjugacy Search Problem relative to a subgroup (SCSP*):** given $u_i, v_i \in G$ and a finitely generated subgroup $A$ of $G$ such that the system $u_i^x = v_i$, $i = 1, \ldots, n$ has a solution in $A$, find a solution.

**Remark 11.1.1.** Observe that if the word problem is decidable in $G$, then all the problems above are also decidable. Indeed, one can enumerate all possible elements $x$ (either in $G$ or in the subgroup $A$), plug them one at a time into the equations and check, using the decision algorithm for the word problem in $G$, whether or not $x$ is a solution. Since the systems above do have solutions, this algorithm will eventually find one. However, the main problem here is not decidability itself, the problem is whether or not one can find a solution sufficiently efficiently, say in polynomial time in the size of the inputs.

The following result is easy.

**Lemma 11.1.2.** *For any group $G$, the AAG problem can be reduced in linear time to the SCSP\*.*

*Proof.* Suppose in a finitely generated group $G$ we are given the public data from the AAG scheme, i.e., we are given the subgroups

$$A = \langle a_1, \ldots, a_m \rangle, \quad B = \langle b_1, \ldots, b_n \rangle,$$

and the elements $\bar{b}_1^a, \ldots, \bar{b}_n^a$ and $\bar{a}_1^b, \ldots, \bar{a}_n^b$. If the SCSP relative to subgroups $A$ and $B$ is decidable in $G$, then by solving the system of equations

$$b_1^x = \bar{b}_1^a, \ldots, b_n^x = \bar{b}_n^a \tag{11.1}$$

in $A$, one can find a solution $u \in A$. Similarly, by solving the system of equations

$$a_1^y = \bar{a}_1^b, \ldots, a_m^y = \bar{a}_m^b \tag{11.2}$$

in $B$, one can find a solution $v \in B$. Note that all solutions of the system (11.1) are elements of the form $ca$, where $c$ is an arbitrary element from the centralizer $C_G(B)$, and all solutions of the system (11.2) are of the form $db$ for some $d \in C_G(A)$. In this case, obviously $[u, v] = [ca, db] = [a, b]$ gives a solution to the AAG problem. $\square$

Clearly, in some groups, for example, in abelian groups, the AAG problem as well as the SCSP\* are both decidable in polynomial time, which makes them (formally) polynomial time equivalent. We will see in Section 11.2.2 that SCSP\* is easy in free groups, too.

It is not clear, in general, whether the SCSP is any harder or easier than the CSP. In hyperbolic groups SCSP, as well as CSP, is easy [18]. There are indications that in finite simple groups, at least generically, the SCSP\* is not harder than the standard CSP (since in those groups two randomly chosen elements generate the whole group). We refer to the preprint [48] for a brief discussion on complexity of these problems.

It would be interesting to get some progress on the following problems, which would shed some light on the complexity of the AAG problem.

**Problem 11.1.3.** 1) In which groups is the AAG problem poly-time equivalent to the SCSP\*?

2) In which groups is the SCSP\* harder than the SCSP?

3) In which groups is the SCSP harder (easier) than the CSP?

In the rest of this chapter we study the hardness of the SCSP\* in various groups and analyze some of the most successful attacks on the AAG scheme from the viewpoint of asymptotic mathematics.

## 11.2   Length-based attacks

The intuitive idea of the length-based attack (LBA) was first put forward in the paper [67] by Hughes and Tannenbaum. In their paper [43], Garber, Kaplan, Teicher, Tsaban, and Vishne gave experimental results suggesting that very large computational power is required for this method to successfully solve the simultaneous conjugacy search problem. In [42], the same authors proposed a variant of the length-based attack that uses memory, and gave experimental results showing that natural types of equations or system of equations in random subgroups of the braid groups can be solved, with high success rates, using the memory-length approach. However, the memory-length attacks were not tried in [42] against the actual parameters used in the AAG protocol. Recently, another variation of the length-based attack for braid groups was developed in [112], which turned out to be very successful against the AAG protocol. The authors of [112] suggested using a heuristic algorithm for approximation of the geodesic length of braids in conjunction with LBA. Furthermore, they analyzed the reasons for success/failure of their variation of the attack, in particular the practical importance of Alice's and Bob's subgroups $A$ and $B$ being non-isometrically embedded and being able to choose the elements of these subgroups distorted in the group (they refer to such elements as "peaks").

In this section we rigorously prove that the same results can be observed in much larger classes of groups. In particular, our analysis works for the class $\mathcal{FB}_{exp}$, and hence for free groups, pure braid groups, locally commutative nonabelian groups, etc.

### 11.2.1   A general description

Since LBA is an attack on the AAG scheme, the inputs for LBA are precisely the inputs for the AAG algorithmic problem, see Section 11.1.2. Moreover, in all its variations LBA attacks the AAG scheme by solving the corresponding conjugacy equations given in a particular instance of the AAG problem. In what follows we take a slightly more general approach and view LBA as a correct partial deterministic search algorithm of a particular type for the simultaneous conjugacy search problem relative to a subgroup (SCSP*) in a given group $G$. In this case LBA is employed to solve the SCSP*, not the AAG problem. Below we describe a basic LBA in its most simplistic form.

Let $G$ be a group with a finite generating set $X$. Suppose we are given a particular instance of the SCSP*, i.e., a system of conjugacy equations $u_i^x = v_i, i = 1, \ldots, m$, which has a solution in a subgroup $A = \langle Y \rangle$ generated by a finite set $Y$ of elements in $G$ (given by words in $F(X)$). The task is to find such a solution in $A$. The main idea of LBA is very simple and it is based on the following assumptions.

**(L1)** for arbitrary "randomly chosen" elements $u, w \in G$, one has $l_X(u^w) > l_X(u)$,

which is convenient sometimes to have in a more general form:

**(L2)** for "randomly chosen" elements $w, y_1, \ldots, y_k$ in $G$, the element $w$ has minimal $l_X$-length among all elements of the form $w^y$, where $y$ runs over the subgroup of $G$ generated by $y_1, \ldots, y_k$.

It is not at all obvious whether these assumptions are actually correct for a given platform group. We will return to these issues in due course; now we just say that at least there has to be an algorithm $\mathcal{A}$ to compute the length $l_X(w)$ for any given element $w \in G$.

Consider Alice's public conjugates $\bar{b}_1^a, \ldots, \bar{b}_n^a$, where $a = a_{s_1}^{\varepsilon_1} \ldots a_{s_L}^{\varepsilon_L}$. Each $\bar{b}_i^a$ is the result of a sequence of conjugations of $b_i$ by generators of $\mathcal{A}$:

$$
\begin{array}{ccc}
 & b_i & \\
 & \downarrow & \\
a_{s_1}^{-\varepsilon_1}\, & b_i & \,a_{s_1}^{\varepsilon_1} \\
 & \downarrow & \\
a_{s_2}^{-\varepsilon_2} a_{s_1}^{-\varepsilon_1}\, & b_i & \,a_{s_1}^{\varepsilon_1} a_{s_2}^{\varepsilon_2} \\
 & \downarrow & \\
 & \cdots & \\
 & \downarrow & \\
\bar{b}_i^a = \quad a_{s_L}^{-\varepsilon_L} \ldots a_{s_2}^{-\varepsilon_2} a_{s_1}^{-\varepsilon_1}\, & b_i & \,a_{s_1}^{\varepsilon_1} a_{s_2}^{\varepsilon_2} \ldots a_{s_L}^{\varepsilon_L}
\end{array}
\tag{11.3}
$$

This conjugating sequence is the same for each $b_i$ and is defined by the private key $a$. The main goal of the attack is to reverse the sequence (11.3), and going back from the bottom to the top recover each conjugating factor. If successful, the procedure will result in the actual conjugator as a product of generators of $\mathcal{A}$.

The next algorithm is the simplest realization of LBA called *the fastest descent LBA*. It takes as an input three tuples $(a_1, \ldots, a_m)$, $(b_1, \ldots, b_n)$, and $(c_1, \ldots, c_n)$, where the last tuple is assumed to be $\bar{b}_1^a, \ldots, \bar{b}_n^a$. The algorithm is a sequence of the following steps:

- **(Initialization)** Put $x = \varepsilon$ (where $\varepsilon$ is the identity).

- **(Main loop)** For each $i = 1, \ldots, m$ and $\varepsilon = \pm 1$ compute $l_{i,\varepsilon} = \sum_{j=1}^{n} l_X(a_i^{-\varepsilon} c_j a_i^{\varepsilon})$.

    - If for each $i = 1, \ldots, n$ and $\varepsilon = \pm 1$ the inequality $l_{i,\varepsilon} > \sum_{j=1}^{n} l_X(c_j)$ is satisfied, then output $x$.

    - Otherwise pick $i$ and $\varepsilon$ giving the smallest value of $l_{i,\varepsilon}$. Multiply $x$ on the right by $a_i^{\varepsilon}$. For each $j = 1, \ldots, n$ conjugate $c_j = a_i^{-\varepsilon} c_j a_i^{\varepsilon}$. Continue.

- **(Last step)** If $c_j = b_j$ for each $j = 1, \ldots, n$, then output the obtained element $x$. Otherwise output *Failure*.

Other variations of LBA suggested in [112] are "LBA with backtracking" and "generalized LBA". We refer to [112] for a detailed discussion on these.

We note that instead of the length function $l_X$, one can use any other objective function satisfying assumptions (L1) and (L2). Here besides $l_X$ we analyze the behavior of modifications of LBA relative to the following functions:

**(M1)** Instead of computing the geodesic length $l_X(v_i)$ of the element $v_i \in G$, compute the geodesic length $l_Z(v_i)$ in the subgroup $H$ generated by $Z = \{u\} \cup Y$ (clearly, $v_i \in H$). In this case, LBA in $G$ is reduced to LBA in $H$, which might be easier. We term $l_Z$ the *inner* length in LBA.

**(M2)** It might be difficult to compute the lengths $l_X(w)$ or $l_Z(w)$. In this case, one can try to compute some "good", say linear, approximations of $l_X(w)$ or $l_Z(w)$, and then use some heuristic algorithms to carry over LBA (see [112]).

These modifications can make LBA much more efficient. In what follows our main interest is in the generic case time complexity of LBA. To formulate this precisely, one needs to describe the set of inputs for LBA and the corresponding distribution on them.

Recall that an input for the SCSP* in a given group $G$ with a fixed finite generating set $X$ consists of a finitely generated subgroup $A = \langle a_1, \ldots, a_k \rangle$ of $G$ given by a $k$-tuple $(a_1, \ldots, a_k) \in F(X)^k$, and a finite system of conjugacy equations $u_i^x = v_i$, where $u_i, v_i \in F(X)$, $i = 1, \ldots, m$, that has a solution in $A$. We denote this data by $\alpha = (T, b)$, where $T = (a_1, \ldots, a_k, u_1, \ldots, u_m)$ and $b = (v_1, \ldots, v_m)$. The distinction that we make here between $T$ and $b$ will be used later on. For fixed positive integers $m, k$ we denote the set of all inputs $\alpha = (T, b)$ as above by $I_{k,m}$.

The standard procedure to generate a "random" input of this type in the AAG protocol is as follows.

**Random generation of inputs for LBA in a given $G$:**

- pick a random $k \in \mathbb{N}$ from a fixed interval $K_0 \leq k \leq K_1$;

- pick randomly $k$ words $a_1, \ldots, a_k \in F(X)$ with the length in a fixed interval $L_0 \leq |w_i| \leq L_1$;

- pick a random $m \in \mathbb{N}$ from a fixed interval $M_0 \leq m \leq M_1$;

- pick randomly $m$ words $u_1, \ldots, u_m \in F(X)$ with the length in a fixed interval $N_0 \leq |u_i| \leq N_1$;

- pick a random element $w$ from the subgroup $A = \langle a_1, \ldots, a_k \rangle$, as a random product $w = a_{i_1} a_{i_2} \ldots a_{i_c}$ of elements from $\{a_1, \ldots, a_k\}$ with the number of factors $c$ in a fixed interval $P_1 \leq c \leq P_2$;

- conjugate $v_i = u_i^w$ and compute the normal form $\tilde{v}_i$ of $v_i$, $i = 1, \ldots, m$.

As we have argued in Section 10.2, one can fix the numbers $k, m$ and the number of factors $c$ in the product $w$ in advance. Observe that the choice of the elements $v_1, \ldots, v_m$ is completely determined by the choice of the tuple $T = (a_1, \ldots, a_k, u_1, \ldots, u_m) \in F(X)^{k+m}$ and the word $w$.

Note also that the distribution on the subgroups $H = \langle T \rangle$ (more precisely, on their descriptions from $F(X)^{k+m}$) that comes from the random generation procedure above coincides with the distribution on the $(k+m)$-generated subgroups that was described in Section 10.2. We summarize these observations in the following remark.

**Remark 11.2.1.** 1) The choice of a tuple $T = (a_1, \ldots, a_k, u_1, \ldots, u_m) \in F(X)^{k+m}$ precisely corresponds to the choice of generators of random subgroups described in Section 10.2.

2) Asymptotic properties of subgroups generated by $T$ precisely correspond to the asymptotic properties of subgroups discussed in Section 10.

## 11.2.2 LBA in free groups

In this section we discuss LBA in free groups. Note that there are fast (quadratic time) algorithms to solve the SCSP*, and hence the AAG problem, in free groups (see Section 11.4.2). However, results on LBA in free groups will serve as a base for us to solve the SCSP* in many other groups.

Let $k$ be a fixed positive natural number. We say that cancellation in a set of words $Y = \{y_1, \ldots, y_k\} \subseteq F(X)^k$ is at most $\lambda$, where $\lambda \in (0, 1/2)$, if for any $u, v \in Y^{\pm 1}$ the amount of cancellation in the product $uv$ is strictly less than $\lambda \min\{l_X(u), l_X(v)\}$, provided $u \neq v^{-1}$ in $F(X)$.

We summarize a couple of well-known facts in the following lemma.

**Lemma 11.2.2.** If the set $Y = \{y_1, \ldots, y_k\}$ satisfies the $\lambda$-condition for some $\lambda \in (0, 1/2)$, then:

- *The set $Y$ is Nielsen reduced. In particular, $Y$ freely generates a free subgroup and any element $w \in \langle Y \rangle$ can be uniquely represented as a reduced word in the generators $Y$ and their inverses.*

- *The membership search problem for a subgroup $\langle Y \rangle$ (see Section 2.3.4 for details) is decidable in linear time.*

- *The geodesic length for elements of a subgroup $\langle Y \rangle$ (see Section 11.3.1 for details) is computable in linear time.*

The following result was proved in [98].

**Theorem 11.2.3.** Let $\lambda \in (0, 1/2)$. The set $S$ of $k$-tuples $(u_1, \ldots, u_k) \in F(X)^k$ satisfying the $\lambda$-condition is exponentially generic, and hence the set of $k$-tuples that are Nielsen reduced in $F(X)$ is exponentially generic, too.

Now we are ready to discuss the generic case complexity of LBA in free groups.

**Theorem 11.2.4.** Let $F(X)$ be the free group with basis $X$. Let $v_1 = u_1^x, \ldots, v_n = u_n^x$ be a system of conjugacy equations in $F(X)$, where $x$ is searched for in the subgroup $A$ generated by $a_1, \ldots, a_m$. Let $Z$ be the tuple $(u_1, \ldots, u_n, a_1, \ldots, a_m)$. Then LBA with respect to the length function $l_Z$ solves the SCSP* in linear time on an exponentially generic set of inputs.

*Proof.* Let $n$ and $m$ be fixed positive integers. Denote by $S$ the set of $(n+m)$-tuples $(u_1, \ldots, u_n, a_1, \ldots, a_m) \in F(X)^{n+m}$ that satisfy the 1/4-condition. It follows from Theorem 11.2.3 that the set $S$ is exponentially generic.

Furthermore, the system of conjugacy equations associated with such a tuple $Z = (u_1, \ldots, u_n, a_1, \ldots, a_m)$ has the form

$$\begin{cases} v_1 = u_1^x, \\ \cdots \\ v_n = u_n^x, \end{cases}$$

where $v_i$ belong to the subgroup $H = \langle Z \rangle$ generated by $Z$, and $x$ is searched for in the same subgroup. By Lemma 11.2.2 one can find expressions for $v_i$ in terms of the generators $Z$ in linear time. Now, since the generators $a_1, \ldots, a_m$ are part of the basis of the subgroup $H$, it follows that LBA relative to $l_Z$ successfully finds a solution $x = w(a_1, \ldots, a_m)$ in linear time.  $\square$

### 11.2.3   LBA in groups from $\mathcal{FB}_{exp}$

The results of the previous section are not very surprising because of the nature of cancellation in free groups. What does look surprising is that LBA works generically in some other groups that seem to be "far" from free groups. In this and the next section we outline a general mathematical explanation of this phenomenon. In particular, it will be clear why the modification (M1) of LBA, which was discussed in Section 11.2.1, is very robust, provided one can compute the geodesic length in subgroups.

We start with a slight generalization of Theorem 11.2.4. Recall (from Section 11.2.1) that inputs for LBA, as well as for the SCSP*, can be described in the form $\alpha = (T, b)$, where $T = (a_1, \ldots, a_k, u_1, \ldots, u_m) \in F(X)^{k+m}$ and $b = (v_1, \ldots, v_m)$, such that there is a solution of the system $u_i^x = v_i$ in the subgroup $A = \langle a_1, \ldots, a_k \rangle$.

**Lemma 11.2.5.** *Let $G$ be a group with a finite generating set $X$, and $I_{k,m}$ the set of all inputs $(T, b)$ for LBA in $G$. Put*

$$I_{free} = \{(T, b) \in I_{k,m} \mid T \text{ freely generates a free subgroup in } G\}.$$

*Suppose there is an exponentially generic subset $S$ of $I_{free}$ and an algorithm $\mathcal{A}$ that computes the geodesic length $l_T$ of elements from the subgroup $\langle T \rangle$, $(T, b) \in S$, when these elements are given as words from $F(X)$. Then there is an exponentially generic subset $S'$ of $I_{free}$ such that LBA halts on inputs from $S'$ and outputs a solution for the related SCSP* in at most quadratic time modulo the algorithm $\mathcal{A}$.*

*Proof.* The result follows directly from Theorem 11.2.4.  $\square$

Let $G \in \mathcal{FB}_{exp}$. In the next theorem we prove that time complexity of the SCSP* on an exponentially generic set of inputs is at most quadratic modulo time complexity of the problem of computing the geodesic length in a finitely generated subgroup of $G$.

**Theorem 11.2.6. (Reducibility to subgroup-length function)** *Let $G$ be a group with exponentially generic free basis property, and $X$ a finite generating set of $G$. Then there is an exponentially generic subset $S$ of the set $I_{k,m}$ of all inputs for LBA in $G$ such that LBA relative to $l_T$ halts on inputs from $S$ and outputs a solution for the related SCSP\*. Moreover, time complexity of LBA on inputs from $S$ is at most quadratic modulo the algorithm $\mathcal{A}$ that computes the geodesic length $l_T$ of elements from the subgroup $\langle T \rangle$ when these elements are given as words from $F(X)$.*

*Proof.* By Lemma 11.2.5 there is an exponentially generic subset $S$ of $I_{free}$ such that LBA halts on inputs from $S$ and outputs a solution for the related SCSP\*. Moreover, time complexity of LBA on inputs from $S$ is at most quadratic modulo the algorithm $\mathcal{A}$ that computes the geodesic length $l_T$ of elements from the subgroup $\langle T \rangle$ when these elements are given as words from $F(X)$. It suffices to show now that the set $I_{free}$ is exponentially generic in the set of all inputs $I$ for LBA in $G$. By Remark 11.2.1, the asymptotic density of the set $I_{free}$ in $I$ is the same as the asymptotic density of the set of tuples $T \in F(X)^{k+m}$ which have free basis property in $G$. Since $G$ is in $\mathcal{FB}_{exp}$, this set is exponentially generic in $F(X)^{k+m}$, so is $I_{free}$ in $I$. This proves the theorem. $\square$

## 11.3 Computing the geodesic length in a subgroup

For groups $G \in \mathcal{FB}_{exp}$, Theorem 11.2.6 reduces (in quadratic time) time complexity of LBA on an exponentially generic set of inputs to time complexity of the problem of computing the geodesic length in finitely generated subgroups of $G$. In this section we discuss time complexity of algorithms to compute the geodesic length in a subgroup of $G$. This discussion is related to the modification (M2) of LBA introduced in Section 11.2.1. In particular, we focus on the situation where we do not have fast algorithms to compute the geodesic length of elements in finitely generated subgroups of $G$, or even in the group $G$ itself. In this case, as was mentioned in the modification (M2), one can try to compute some linear approximations of these lengths and then use heuristic algorithms to carry out LBA.

In Section 11.3.2 we discuss hardness of the problem of computing the geodesic length (GLP) in braid groups $B_n$, the original platforms of the AAG protocol. Time complexity of the GLP in $B_n$ relative to the standard set $\Sigma$ of Artin generators is unknown. We discuss some recent results and conjectures in this area. However, there are efficient linear approximations of the geodesic length in $B_n$ relative to the set $\Delta$ of generators (the set of generalized half-twists). Theoretically, this gives linear approximations of the geodesic length of elements in $B_n$ in the Artin generators and, furthermore, linear approximations of the geodesic inner length in quasi-isometrically embedded subgroups. If, as conjectured, the set of quasi-isometrically embedded subgroups is exponentially generic in braid groups, then this gives a sound foundation for LBA in braid groups. Note that linear approximations alone are not quite sufficient for successful LBA. To get a precise solution of the SCSP*, one also needs a robust "local search" near a given approximation of the solution. To this end several efficient heuristic algorithms have been developed, see e.g., [102], [112]. However, so far none of them exploited directly the interesting interplay between geodesic lengths in $\Sigma$ and $\Delta$, as well as quasi-isometric embeddings of subgroups.

## 11.3.1   Related algorithmic problems

We start with precise formulation of some problems related to computing geodesics in $G$.

**Computing the geodesic length in a group (GLP):** Let $G$ be a group with a finite generating set $X$. Given an element $w \in G$ as a product of generators from $X$, compute the geodesic length $l_X(w)$.

**Computing the geodesic length in a subgroup (GLSP):** Let $G$ be a group with a finite generating set $X$, and $A$ a subgroup of $G$ generated by a finite set of elements $Y = \{a_1, \ldots, a_k\}$ of $G$ given as words from $F(X)$. Given an element $w \in A$ as a product of generators of $A$, compute the geodesic length $l_Y(w)$.

There is another (harder) variation of this problem that comes from the SCSP* problem:

**Computing the geodesic length in a subgroup (GLSP*):** Let $G$ be a group with a finite generating set $X$, and $A$ a subgroup of $G$ generated by a finite set of elements $Y = \{a_1, \ldots, a_k\}$ of $G$ given as words from $F(X)$. Given an element $w \in A$ as a word from $F(X)$, compute the geodesic length $l_Y(w)$.

The following lemma is obvious. Recall that the membership search problem (MSP) for a subgroup $A$ in $G$ asks for a given element $w \in F(X)$, which belongs to $A$, to find a decomposition of $w$ into a product of generators from $Y$ and their inverses.

**Lemma 11.3.1.** *Let $G$ be a finitely generated group and $A$ a finitely generated subgroup of $G$. Then:*

1) *The GLSP is linear time reducible to the GLSP*;*

2) *The GLSP* is linear time reducible to the GLSP modulo the MSP in $A$.*

Observe that if the GLSP has a fast solution for $A = G$ in $G$, then there is a fast algorithm to find the geodesic length of elements of $G$ with respect to $X$. In particular, the word problem in $G$ has a fast decision algorithm. In some groups, like free groups or partially commutative groups, given by standard generating sets, there are fast algorithms for computing the geodesic length of elements. In many other groups, like braid groups, or nilpotent groups, the computation of the geodesic length of elements is hard. Nevertheless, in many applications, including cryptography, it suffices to have a fast algorithm to compute a reasonable, say linear, approximation of the geodesic length of a given element. This motivates the following problem.

**Computing a linear approximation of the geodesic length in a group (AGL):** Let $G$ be a group with a finite generating set $X$. Given a word $w \in F(X)$ compute a linear approximation of the geodesic length of $w$. More precisely, find an algorithm that for $w \in F(X)$ outputs a word $w' \in F(X)$ such that $\lambda l_X(w) + c \geq l_X(w')$, where $\lambda$ and $c$ are independent of $w$.

Another problem is to compute a good approximation in a subgroup of a group.

**Computing a linear approximation of the geodesic length in a subgroup (AGLS):** Let $G$ be a group with a finite generating set $X$, and $A$ a subgroup of $G$ generated by a finite set of elements $Y = \{a_1, \dots, a_k\}$ of $G$ given as words from $F(X)$. Given an element $w \in A$ as a word from $F(X)$, compute a linear approximation of the geodesic length $l_Y(w)$ of $w$.

Assume now that there is a fast algorithm to solve the AGL problem in the group $G$. This does not imply that there is a fast algorithm to compute a linear approximation of the geodesic length in a given subgroup $A$ of $G$, unless the subgroup $A$ is quasi-isometrically embedded in $G$.

**Lemma 11.3.2.** *Let $G$ be a group with a finite generating set $X$, and $\mathcal{A}$ an algorithm that solves the AGL problem in $G$ with respect to $X$. If $H$ is a quasi-isometrically embedded subgroup of $G$ generated by a finite set $Y$, then for every $w \in H$ given as a word from $F(X)$, the algorithm $\mathcal{A}$ outputs a word $w' \in F(X)$ such that $l_Y(w) \leq \mu l_X(w') + d$ for some constants $\mu$ and $d$, which depend only on $\mathcal{A}$ and $H$.*

*Proof.* The proof is straightforward. $\square$

## 11.3.2 Geodesic length in braid groups

There are no known efficient algorithms to compute the geodesic length of elements in braid groups with respect to the set $\Sigma$ of the standard Artin's generators. Some indications that this could be a hard problem are given in [115], where the authors prove that the set of geodesics in $B_\infty$ is co-**NP**-complete. However, in a given group, the problem of computing the length of a word could be easier than the problem of finding a geodesic of the word. Moreover, complexity of a set of geodesics in a group may not be a good indicator of the time complexity of computing the geodesic length in a randomly chosen subgroup. In fact, it has been shown in [102, 103] that in a braid group $B_n$ one can efficiently compute a reasonable approximation of the length function on $B_n$ (relative to $\Sigma$), which gives a foundation for successful LBA, without computing the length in the group. Furthermore, there are interesting conjectures that, if settled affirmatively, will lead to more efficient algorithms for computing the length of elements in braid groups and their subgroups. To explain this we need to first recall some known facts and terminology.

We remind the reader that the group $B_n$ has the standard Artin's presentation:

$$B_n = \left\langle\ \sigma_1, \dots, \sigma_{n-1}\ \left|\ \begin{array}{ll} \sigma_i\sigma_j\sigma_i = \sigma_j\sigma_i\sigma_j & \text{if } |i-j| = 1 \\ \sigma_i\sigma_j = \sigma_j\sigma_i & \text{if } |i-j| > 1 \end{array}\ \right\rangle\right..$$

We denote the corresponding generating set $\{\sigma_1, \dots, \sigma_{n-1}\}$ by $\Sigma$ and the corresponding length function by $l_\Sigma(w)$.

Elements in $B_n$ admit so-called *Garside normal forms* (cf. our Section 5.1.3). These forms are unique and the time complexity to compute the Garside normal form of an element of $B_n$ given by a word $w \in F(\Sigma)$ is bounded by $O(|w|^2 n^2)$. However, Garside normal forms are far from being geodesic in $B_n$.

In 1991 Dehornoy introduced [28] the following notion of $\sigma$-positive braid word and a handle-reduction algorithm to compute a $\sigma$-positive representative of a given word. A braid word $w$ is termed to be $\sigma_k$-positive (respectively, $\sigma_k$-negative), if it contains $\sigma_k$, but does not contain $\sigma_k^{-1}$ or $\sigma_i^{\pm 1}$ with $i < k$ (respectively, contains $\sigma_k^{-1}$, but not $\sigma_k$ or $\sigma_i^{\pm 1}$ with $i < k$). A braid word $w$ is said to be $\sigma$-positive (respectively, $\sigma$-negative), if it is $\sigma_k$-positive (respectively, $\sigma_k$-negative) for some $k \leq n - 1$. A braid word $w$ is said to be $\sigma$-consistent if it is either trivial, or $\sigma$-positive, or $\sigma$-negative.

**Theorem 11.3.3 (Dehornoy [28]).** . *For any braid $\beta \in B_n$, exactly one of the following is true:*

1) *$\beta$ is trivial;*

2) *$\beta$ can be presented by a $\sigma_k$-positive braid word for some $k$;*

3) *$\beta$ can be presented by a $\sigma_k$-negative braid word for some $k$.*

Thus, it makes sense to speak about $\sigma$-positive and $\sigma_k$-positive (or $\sigma$-, $\sigma_k$-negative) braids.

The following question is of primary interest when solving the AGL problem in braid groups: is there a polynomial $p(x)$ such that for every word $w \in F(\Sigma)$, $p(l_\Sigma(w))$ gives an upper bound for the $\Sigma$-length of the shortest $\sigma$-consistent braid word representing $w \in B_n$? Dehornoy's original algorithms in [28], as well as the handle reduction from [29]) and the algorithm from [39], give an exponential bound on the length of the shortest $\sigma$-consistent representative.

In [34] (see also [29, 39] for a related discussion) Dynnikov and Wiest formulated the following

**Conjecture 11.3.4.** There are numbers $\lambda, c$ such that every braid $w \in B_n$ has a $\sigma$-consistent representative whose $\Sigma$-length is bounded linearly by the $\Sigma$-length of the braid.

They also showed that this conjecture holds if the $\Sigma$-length of elements is replaced by the $\Delta$-length (relative to the set $\Delta$ of generators).

The set $\Delta$ of generators consists of the braids $\Delta_{ij}, 1 \leq i < j \leq n$, which are the half-twists of strands $i$ through $j$:

$$\Delta_{ij} = (\sigma_i \ldots \sigma_{j-1})(\sigma_i \ldots \sigma_{j-2}) \ldots \sigma_i.$$

$\Delta$ is a generating set of $B_n$, containing the Artin generators $\sigma_i = \Delta_{i,i+1}$ and the Garside fundamental braid $\Delta_{1n}$. The *compressed* $\Delta$-length of a word $w$ of the form

$$w = \Delta_{i_1 j_1}^{k_1} \ldots \Delta_{i_s j_s}^{k_s},$$

where $k_t \neq 0$ and $\Delta_{i_t, j_t} \neq \Delta_{i_{t+1}, j_{t+1}}$ for all $t$, is defined by

$$L_\Delta(w) = \Sigma_{i=1} \log_2(|k_i| + 1).$$

For an element $\beta \in B_n$, the value $L_\Delta(\beta)$ is defined by

$$L_\Delta(\beta) = \min\{L_\Delta(w) \mid \text{ the word } w \text{ represents } \beta\}.$$

Obviously, for any braid $\beta$ one has

$$L_\Delta(\beta) \leq l_\Delta(\beta) \leq l_\Sigma(\beta).$$

The modified conjecture assumes the following extension of the notion of a $\sigma$-positive braid word: a word in the alphabet $\Delta = \{\Delta_{ij} \mid 0 < i < j < n\}$ is said to be $\sigma$-positive if, for some $k < l$, it contains $\Delta_{kl}$, and contains neither $\Delta_{kj}^{-1}$ nor $\Delta_{ij}^{\pm 1}$ with $i < k$ and any $j$. In other words, a word $w$ in letters $\Delta_{ij}$ is $\sigma$-positive (negative) if the word in standard generators $\sigma_i$ obtained from $w$ by the obvious expansion is.

**Theorem 11.3.5 (Dynnikov, Wiest [34]).** *Any braid $\beta \in B_n$ can be presented by a $\sigma$-consistent word $w$ in the alphabet $\{\Delta_{ij}\}$ such that*

$$l_\Delta(w) \leq 30nl_\Delta(\beta).$$

This theorem gives a method to approximate geodesic length in braid groups, as well as in their quasi-isometrically embedded subgroups. It remains to be seen whether or not this could lead to more efficient versions of LBA.

## 11.4   Quotient attacks

In this section we describe another type of attacks, called *quotient attacks* (QA) (cf. our Section 6.1.6). In fact, quotient attacks are just fast generic algorithms to solve some search problems in groups, such as the Membership Search Problem (MSP), the Simultaneous Conjugacy Search Problem (SCSP), the Simultaneous

Conjugacy Search Problem relative to a subgroup (SCSP*), etc. The main idea
behind QA is that to solve a problem in a group $G$ it suffices, on most inputs, to
solve it in a quotient $G/N$, provided $G/N$ has the generic free basis property and
a fast decision algorithm for the relevant problem. In particular, this is the case if
$G$ has a free nonabelian quotient. Note that a similar idea was already exploited
in [75], but there the answer was given only for inputs in the "No" part of the
decision problem, which obviously is of no use for search problems. The strength
of our approach comes from the extra requirement that $G/N$ has the free basis
property.

In Sections 11.4.1 and 11.4.2 we discuss the conjugacy and membership problems (in all their variations) in free groups. Some of these results were known as
"folklore", some could be found in the literature. Nevertheless, we sketch most of
the proofs here, since this will serve us as the ground for solving similar problems
in other groups.

## 11.4.1   Membership problems in free groups

In this section we discuss some algorithms to solve different versions of the membership problem in free groups. We start by recalling the classical membership
problem (MP). Everywhere below $G$ is a fixed group generated by a finite set $X$.

**The Membership Problem (MP):** Let $A = \langle a_1, \ldots, a_m \rangle$ be a fixed finitely generated subgroup of $G$ given by a finite set of generators $a_1, \ldots, a_m$ (viewed as words
in $F(X)$). Given a word $w \in F(X)$, decide whether or not $w$ belongs to $A$.

When the subgroup $A$ is not fixed, but comes as part of the input (as in AAG
scheme), the problem is more adequately described in its *uniform* version.

**The Uniform Membership Problem (UMP):** Given a finite tuple of elements
$w, a_1, \ldots, a_m \in F(X)$, decide whether or not $w$ (viewed as an element of $G$)
belongs to the subgroup $A$ generated by the elements $a_1, \ldots, a_m$ in $G$.

To solve the MP in free groups, we use the folding technique introduced
by Stallings [132] (see also [74] for a more detailed treatment). Given a tuple of
words $a_1, \ldots, a_m \in F(X)$ one can construct a finite deterministic automaton $\Gamma_A$,
which accepts a reduced word $w \in F(X)$ if and only if $w$ belongs to the subgroup
$A = \langle a_1, \ldots, a_m \rangle$ generated by $a_1, \ldots, a_m$ in $F(X)$.

To determine the time complexity of the MP and UMP, recall that for a given
positive integer $n$, the function $\log_2^* n$ is defined as the least natural number $m$
such that the $m$-tower of exponents of 2 exceeds $n$, or equivalently, $\log_2 \circ \log_2 \circ \ldots \circ$
$\log_2(n) \leq 1$, where on the left one has the composition of $m$ logarithms.

**Lemma 11.4.1.** *There is an algorithm that for any input $w, a_1, \ldots, a_m \in F(X)$ for
the UMP finds the correct answer in nearly linear time $O(|w| + n \log^* n)$, where*

$n = \sum_{i=1}^{k} |a_i|$. *Furthermore, the algorithm works in linear time* $O(|w| + n)$ *on an exponentially generic set of inputs.*

*Proof.* Indeed, given $w, a_1, \ldots, a_m \in F(X)$ one can construct $\Gamma_A$ in time $O(n \log^* n)$ (see [137]) and check whether or not $\Gamma_A$ accepts $w$ in time $O(|w|)$, as required.

To prove the generic estimate recall that the set of $m$-tuples $a_1, \ldots, a_m \in F(X)$ satisfying the 1/4-condition is exponentially generic, and the Stallings procedure gives the automaton $\Gamma_A$ in linear time $O(n)$. $\qquad\square$

In cryptography, the search versions of the MP and UMP are most interesting.

**The Membership Search Problem (MSP):** Let $A = \langle a_1, \ldots, a_m \rangle$ be a fixed finitely generated subgroup of $G$ given by a finite set of generators $a_1, \ldots, a_m$, viewed as words in $F(X)$. Given a word $w \in F(X)$ that belongs to $A$, find a representation of $w$ as a product of the generators $a_1, \ldots, a_m$ and their inverses.

**The Uniform Membership Search Problem (UMSP):** Given a finite set of elements $w, a_1, \ldots, a_m \in F(X)$ such that $w \in A = \langle a_1, \ldots, a_m \rangle$, find a representation of $w$ as a product of the generators $a_1, \ldots, a_m$ and their inverses.

Upper bounds for the time complexity of the MSP follow easily from the corresponding bounds for the MP.

**Lemma 11.4.2.** *The time complexity of the MSP in a free group is bounded from above by* $O(|w|)$.

*Proof.* Let $A = \langle a_1, \ldots, a_m \rangle$ be a fixed finitely generated subgroup of $G$. As we mentioned above, one can construct the Stallings folding $\Gamma_A$ in time $O(n \log^* n)$, where $n = |a_1| + \ldots + |a_n|$. Then, one can construct, in linear time in $n$, a Nielsen basis $S = \{b_1, \ldots, b_n\}$ for $A$ by using the breadth-first search (see [74]). Now, given a word $w \in F(X)$ that belongs to $A$, one can follow the accepting path for $w$ in $\Gamma_A$ and rewrite $w$ as a product of generators from $S$ and their inverses. This requires linear time in $|w|$. It suffices to notice that the elements $b_i$ can be expressed as fixed products of elements from the initial generating set of $A$, i.e., $b_i = u_i(a_1, \ldots, a_n)$, $i = 1, \ldots, m$. Therefore, any expression of $w$ as a product of elements from $S^{\pm 1}$ can be rewritten in linear time into a product of the initial generators. $\qquad\square$

Note that in the proof above we used the fact that any product of new generators $b_i$ and their inversions can be rewritten in linear time into a product of the old generators $a_i$ and their inversions. This was because we assumed that one can rewrite the new generators $b_i$ as products of the old generators $a_i$ in a constant time. This is correct if the subgroup $A$ is fixed. Otherwise, say in the UMSP, the assumption does not hold. It is not even clear whether one can do it in polynomial time or not. In fact, the time complexity of the UMSP is unknown. The following problem is of great interest in this area.

**Problem 11.4.3.** Is time complexity of the UMSP in free groups polynomial?

However, the generic case complexity of the UMSP in free groups is known.

**Lemma 11.4.4.** *The generic case time complexity of the UMSP in free groups is linear. More precisely, there is an exponentially generic subset $T \subseteq F(X)^n$ such that for every tuple $(w, a_1, \ldots, a_m) \in F(X) \times T$ with $w \in \langle a_1, \ldots, a_m \rangle$, one can express $w$ as a product of $a_1, \ldots, a_m$ and their inverses in time $O(|w| + n)$, where $n = |a_1| + \ldots + |a_n|$.*

*Proof.* First, note that if in the argument in the proof of Lemma 11.4.2 the initial set of generators $a_1, \ldots, a_m$ of a subgroup $A$ satisfies the 1/4-condition, then the set of the new generators $b_1, \ldots, b_m$ coincides with the set of the initial generators (see [74] for details). Moreover, as we mentioned in the proof of Theorem 11.2.4, the set $T$ of tuples $(a_1, \ldots, a_m) \in F(X)^m$ satisfying the 1/4-condition is exponentially generic. Hence the argument from Lemma 11.4.2 proves the required upper bound for the UMSP on $T$. $\qquad\square$

## 11.4.2   Conjugacy problems in free groups

Now we turn to the conjugacy problems in free groups. Again, everywhere below $G$ is a fixed group generated by a finite set $X$.

It is easy to see that the CP and CSP in free groups are decidable in at most quadratic time. It is quite tricky though to show that the CP and CSP are decidable in free groups in linear time. This result is based on the Knuth-Morris-Pratt substring searching algorithm [83]. Similarly, the *root search problem* (see below) is decidable in free groups in linear time.

**The Root Search Problem (RSP):** Given a word $w \in F(X)$, find a shortest word $u \in F(X)$ such that $w = u^n$ for some positive integer $n$.

Note that the RSP in free groups can be interpreted as a problem of finding a generator of the centralizer of a nontrivial element (in a free group, such a centralizer is cyclic).

**Theorem 11.4.5.** *The Simultaneous Conjugacy Problem (SCP) and Simultaneous Conjugacy Search Problem (SCSP) in free groups are reducible in linear time to the CP, CSP, and RSP . In particular, it is decidable in linear time.*

*Proof.* We briefly outline an algorithm that simultaneously solves the problems SCP and SCSP in free groups, i.e., given a finite system of conjugacy equations

$$\begin{cases} u_1^x = v_1, \\ \ldots \\ u_n^x = v_n, \end{cases} \qquad (11.4)$$

the algorithm decides whether or not this system has a solution in a free group $F(X)$, and if so, finds a solution. Using the decision algorithm for the CP one can check whether or not there is an equation in (11.4) that does not have solutions in $F$. If there is such an equation, the whole system does not have solutions in $F$, and we are done. Otherwise, using the algorithm to solve the CSP in $F$, one can find a particular solution $d_i$ for every equation $u_i^x = v_i$ in (11.4). In this case the set of all solutions of the equation $u_i^x = v_i$ coincides with the coset $C(u_i)d_i$ of the centralizer $C(u_i)$. Observe that using the decision algorithm for the RSP, one can find a generator (the root of $u_i$) of the centralizer $C(u_i)$ in $F$.

Consider now the first two equations in (11.4). The system

$$u_1^x = v_1, u_2^x = v_2 \qquad (11.5)$$

has a solution in $F(X)$ if and only if the intersection $V = C(u_1)d_1 \cap C(u_2)d_2$ is nonempty. In this case

$$V = C(u_1)d_1 \cap C(u_2)d_2 = (C(u_1) \cap C(u_2))\, d$$

for some $d \in F$.

If $[u_1, u_2] = 1$, then $V$, as the intersection of two cosets, is nontrivial if and only if the cosets coincide, i.e., $[u_1, d_1d_2^{-1}] = 1$. This can be checked in linear time (since the word problem in $F(X)$ has a linear time solution). Therefore, in linear time we either check that the system (11.5), and hence the system (11.4), does not have any solutions, or we confirm that (11.5) is equivalent to one of the equations, so (11.4) is equivalent to its own subsystem, where the first equation is removed. In the latter case the induction finishes the proof.

If $[u_1, u_2] \neq 1$, then $C(u_1) \cap C(u_2) = 1$, so either $V = \emptyset$ or $V = \{d\}$. In both cases one can easily find all solutions of (11.4). Indeed, if $V = \emptyset$, then (11.4) does not have any solutions. If $V = \{d\}$, then $d$ is the only potential solution of (11.4), and one can check whether or not $d$ satisfies all other equations in (11.4) in linear time by the direct substitution.

Now the problem is to verify, in linear time, whether $V = \emptyset$ or not, which is equivalent to solving an equation

$$u_1^m d_1 = u_2^k d_2 \qquad (11.6)$$

for some integers $m, k$. By finding, in linear time, cyclically reduced decompositions of $u_1$ and $u_2$, one can rewrite the equation (11.6) into an equivalent one of the form

$$w_2^{-k} c w_1^m = b, \qquad (11.7)$$

where $w_1, w_2$ are cyclically reduced forms of $u_1, u_2$, and either $w_2^{-1}c$ or $cw_1$ (or both) are reduced as written, and $b$ does not begin with $w_2^{-1}$ and does not end with $w_1$. Again in linear time, one can find the maximal possible cancellation in $w_2^{-k}c$ and in $cw_1$, and rewrite (11.7) in the form

$$w_2^{-k} \tilde{w}_1^s = \tilde{b}, \qquad (11.8)$$

where $\tilde{w}_1$ is a cyclic permutation of $w_1$, and $|\tilde{b}| \leq |b| + |w_1|$. Note that two cyclically reduced periodic words $w_2, \tilde{w}_1$ either commute or do not have a common subword of length exceeding $|w_2| + |\tilde{w}_1|$. If they commute, then the equation (11.8) becomes a power equation, which is easy to solve. Otherwise, executing (in linear time) possible cancellation on the left-hand side of (11.8), one arrives at an equation of the form

$$w_2^{-r} e \tilde{w}_1^t = \tilde{b}, \tag{11.9}$$

where there is no cancellation at all. This equation can be easily solved for $r$ and $t$. This completes the proof. $\qquad\square$

As we have seen in the proof of Theorem 11.4.5, one of the main difficulties in solving the SCSP in groups lies in computing the intersection of two finitely generated subgroups or their cosets. Note that finitely generated subgroups of $F(X)$ are regular sets (accepted by their Stallings' automata). It is well known in the language theory that the intersection of two regular sets is again regular, and one can find an automaton accepting the intersection in at most quadratic time. This yields the following corollary.

**Corollary 11.4.6.** *The SCSP\* in free groups is decidable in at most quadratic time.*

*Proof.* Recall from the proof of Theorem 11.4.5 that the algorithm solving a finite system of conjugacy equations in a free group either decides that there is no solution to the system, or produces a unique solution, or gives the whole solution set as a coset $Cd$ of some centralizer $C$. In the first case, the corresponding SCSP\* has no solutions in a given finitely generated subgroup $A$. In the second case, given a unique solution $w$ of the system one can construct the automaton $\Gamma_A$ that accepts $A$, and check whether $w$ is in $A$ or not (it takes time $n \log^* n$). In the third case, one needs to check if the intersection $Cd \cap A$ is empty or not; this can be done in at most quadratic time, as we have mentioned before. $\qquad\square$

Observe from the proof of Corollary 11.4.6 that the most time-consuming case in solving the SCSP\* in free groups occurs when all the elements $u_1, \ldots, u_n$ in the system (11.4) commute pairwise. The set of such inputs for the SCSP\* is obviously exponentially negligible. On the other hand, as we have shown in Theorem 11.2.4, LBA relative to $l_T$ solves the SCSP\* in linear time on an exponentially generic subset of a free group. Thus, generically, the SCSP\* in free groups is decidable in linear time.

Since the AAG problem (see Section 11.1.2) is reducible in linear time to the SCSP\* (Lemma 11.1.2), we have the following

**Corollary 11.4.7.** *In an arbitrary free group $F$:*

1) *The AAG problem in $F$ is decidable in at most quadratic time in the size of the input (i.e., in the size of the public information in the AAG scheme).*

2) *The AAG problem in $F$ is decidable in linear time on an exponentially generic set of inputs.*

### 11.4.3 The MSP and SCSP* problems in groups with "good" quotients

In this section we discuss generic complexity of the Membership Search Problem (MSP) and the Simultaneous Conjugacy Search Problem relative to a subgroup (SCSP*) in groups that have "good" factors in $\mathcal{FB}_{exp}$.

Let $G$ be a group generated by a finite set $X$, $G/N$ a quotient of $G$, and $\varphi : G \to G/N$ the canonical epimorphism. Let $H = \langle u_1, \ldots, u_k \rangle$ be a finitely generated subgroup of $G$. To solve the membership search problem for $H$, one can employ the following simple heuristic idea which we formulate as an algorithm.

**Algorithm 11.4.8. (Heuristic solution to the MSP)**
INPUT: A word $w = w(X)$ and generators $\{u_1, \ldots, u_k\} \subset F(X)$ of a subgroup $H$.
OUTPUT: A representation $W(u_1, \ldots, u_k)$ of $w$ as an element of $H$ or *Failure*.
COMPUTATIONS:

A. Compute the generators $u_1^\varphi, \ldots, u_k^\varphi$ of $H^\varphi$ in $G/N$, where $\varphi : G \to G/N$ is the canonical epimorphism.

B. Compute $w^\varphi$, solve the MSP for $w^\varphi$ in $H^\varphi$, and find a representation $W(u_1^\varphi, \ldots, u_k^\varphi)$ of $w^\varphi$ as a product of the generators of $u_1^\varphi, \ldots, u_k^\varphi$ and their inverses.

C. Check if $W(u_1, \ldots, u_k)$ is equal to $w$ in $G$. If this is the case, then output $W$. Otherwise output *Failure*.

Observe that to run Algorithm 11.4.8, one needs to be able to solve the MSP in the quotient $G/N$ (Step B) and to check the result in the original group (Step C), i.e., to solve the word problem in $G$. If these conditions are satisfied, then Algorithm 11.4.8 is a partial deterministic correct algorithm, i.e., if it gives an answer, it is correct. However, it is far from being obvious, even if the conditions are satisfied, that this algorithm can be robust in any interesting class of groups. The next theorem, which is the main result of this section, states that Algorithm 11.4.8 is very robust for groups from $\mathcal{FB}_{exp}$, with a few additional requirements.

**Theorem 11.4.9 (Reduction to a quotient).** *Let $G$ be a group generated by a finite set $X$, and with the word problem in a complexity class $C_1(n)$. Suppose $G/N$ is a quotient of $G$ such that:*

1) *$G/N \in \mathcal{FB}_{exp}$.*

2) *The canonical epimorphism $\varphi : G \to G/N$ is computable in time $C_2(n)$.*

3) *For every $k \in \mathbb{N}$, there is an algorithm $\mathcal{A}_k$ in a complexity class $C_3(n)$, which solves the membership search problem in $G/N$ for an exponentially generic set $M_k \subseteq F(X)^k$ of descriptions of $k$-generated subgroups in $G/N$.*

*Then, for every $k$, Algorithm 11.4.8 solves the membership search problem on an exponentially generic set $T_k \subseteq F(X)^k$ of descriptions of $k$-generated subgroups in*

$G$. Furthermore, Algorithm 11.4.8 belongs to the complexity class $C_1(n) + C_2(n) + C_3(n)$.

*Proof.* We need to show that Algorithm 11.4.8 successfully halts on an exponentially generic set of tuples from $F(X)^k$. By the conditions of the theorem, the set $S_k$ of all $k$-tuples from $F(X)^k$ whose images in $G/N$ freely generate free subgroups, is exponentially generic, as well as the set $M_k$ of all tuples from $F(X)^k$ where the algorithm $\mathcal{A}_k$ applies. Hence the intersection $T_k = S_k \cap M_k$ is exponentially generic in $F(X)^k$. We claim that Algorithm 11.4.8 applies to subgroups with descriptions from $T_k$. Indeed, the algorithm $\mathcal{A}_k$ applies to subgroups generated by tuples $Y = (u_1, \ldots, u_k)$ from $T_k$, so if $w^\varphi \in H^\varphi = \langle Y^\varphi \rangle$, then $\mathcal{A}_k$ outputs a required representation $w^\varphi = W(Y^\varphi)$ in $G/N$. Note that $H^\varphi$ is freely generated by $Y^\varphi$ since $Y \in S_k$; therefore, $\varphi$ is injective on $H$. It follows that $w = W(Y)$ in $G$, as required. This completes the proof. $\qquad\square$

Theorems 11.2.6 and 11.4.9 yield the following

**Corollary 11.4.10.** *Let $G$ be as in Theorem 11.4.9. Then, for every $k, m > 0$, there is an algorithm $\mathcal{C}_{k,m}$ that solves the SCSP\* on an exponentially generic subset of the set $I_{k,m}$ of all possible inputs. Furthermore, $\mathcal{C}_{k,m}$ belongs to the complexity class $n^2 + C_1(n) + C_2(n) + C_3(n)$.*

**Corollary 11.4.11.** *Let $G$ be a group of pure braids $PB_n$, $n \geq 3$, or a nonabelian partially commutative group $G(\Gamma)$. Then, for every $k, m > 0$, there is an algorithm $\mathcal{C}_{k,m}$ that solves the SCSP\* on an exponentially generic subset of the set $I_{k,m}$ of all possible inputs. Furthermore, $\mathcal{C}_{k,m}$ belongs to the complexity class $O(n^2)$.*

*Proof.* Recall that in any pure braid group and in any nonabelian partially commutative group the word problem can be solved by a quadratic time algorithm. Now the statement follows from Corollary 11.4.10 and Corollaries 10.4.3 and 10.4.5. $\quad\square$

# Bibliography

[1] I. Anshel, M. Anshel, and D. Goldfeld, *An algebraic method for public-key cryptography*, Math. Res. Lett. 6 (1999), pp. 287–291.

[2] I. Anshel, M. Anshel, D. Goldfeld, and S. Lemieux, *Key Agreement, The Algebraic Eraser$^{TM}$, and Lightweight Cryptography*. Algebraic Methods in Cryptography, Contemporary Mathematics 418, pp. 1–34. American Mathematical Society, 2006.

[3] K. Appel and P. Schupp, *Artin groups and infinite Coxeter groups*, Invent. Math. 72 (1983), pp. 201–220.

[4] G. Arzhantseva and A. Olshanskii, *Genericity of the class of groups in which subgroups with a lesser number of generators are free*, (Russian) Mat. Zametki 59 (1996), pp. 489–496.

[5] G. Baumslag, A. G. Myasnikov, and V. Shpilrain, *Open problems in combinatorial group theory. Second Edition.* Combinatorial and geometric group theory, Contemporary Mathematics 296, pp. 1–38. American Mathematical Society, 2002.

[6] G. Baumslag, R. Strebel, and M. W. Thomson, *On the multiplicator of $F/\gamma_c R$*, J. Pure Appl. Algebra 16 (1980), pp. 121–132.

[7] S. Ben David, B. Chor, O. Goldreich, and M. Luby, *On the theory of average case complexity*, J. Comput. Syst. Sci. 44 (1992), pp. 193–219.

[8] S. Bigelow, *Braid groups are linear*, J. Amer. Math. Soc. 14 (2001), pp. 471–486.

[9] J.-C. Birget, S. Magliveras, and M. Sramka, *On public-key cryptosystems based on combinatorial group theory*, Tatra Mountains Mathematical Publications 33 (2006), pp. 137–148.

[10] J. Birman, *Braids, Links and Mapping Class Groups*, Annals of Math. Studies. Princeton University Press, 1974.

[11] J. S. Birman, V. Gebhardt, and J. Gonzalez-Meneses, *Conjugacy in Garside groups I: Cyclings, powers, and rigidity*, Groups, Geometry, and Dynamics 1 (2007), pp. 221–279.

[12] ———, *Conjugacy in Garside Groups III: Periodic braids*, J. Algebra 316 (2007), pp. 746–776.

[13] ———, *Conjugacy in Garside groups II: Structure of the ultra summit set*, Groups, Geometry, and Dynamics 2 (2008), pp. 16–31.

[14] A. Bogdanov and L. Trevisan, *Average-Case Complexity*, Foundations and Trends in Theoretical Computer Science. Now Publishers Inc, 2006.

[15] O. Bogopolski, A. Martino, O. Maslakova, and E. Ventura, *Free-by-cyclic groups have solvable conjugacy problem*, Bull. Lond. Math. Soc. 38 (2006), pp. 787–794.

[16] A. Borovik, A. Myasnikov, and V. Shpilrain, *Measuring sets in infinite groups*. Computational and Statistical Group Theory, Contemporary Mathematics 298, pp. 21–42. American Mathematical Society, 2002.

[17] A. V. Borovik, A. G. Myasnikov, and V. N. Remeslennikov, *Multiplicative measures on free groups*, Internat. J. Algebra and Comput. 13 (2003), pp. 705–731.

[18] M. Bridson and J. Howie, *Conjugacy of finite subsets in hyperbolic groups*, Internat. J. Algebra and Comput. 15 (2005), pp. 725–756.

[19] M. Brin and C. Squier, *Groups of Piecewise Linear Homomorphisms of the Real Line*, Invent. Math. 79 (1985), pp. 485–498.

[20] J. W. Cannon, W. J. Floyd, and W. R. Parry, *Introductory notes on Richard Thompson's groups*, L'Enseignement Mathematique 42 (1996), pp. 215–256.

[21] S. Cleary, M. Elder, A. Rechnitzer, and Taback J., *Random subgroups of Thompson's group F*, preprint. Available at http://arxiv.org/abs/0711.1343.

[22] C. Coarfa, D. D. Demopoulos, A. S. M. Aguirre, D. Subramanian, and M. Y. Vardi, *Random 3-SAT: The Plot Thickens*. Proceedings of the International Conference on Constraint Programming, pp. 143–159, 2000.

[23] D. Collins, *Relations among the squares of the generators of the braid group*, Invent. Math. 117 (1994), pp. 525–529.

[24] S. A. Cook and D. G. Mitchell, *Finding Hard Instances of the Satisfiability Problem: A Survey*, Satisfiability Problem: Theory and Applications, 35, American Mathematical Society, 1997, pp. 1–17.

[25] Cryptography and braid groups, http://www.adastral.ucl.ac.uk/ helger/crypto/link/public/braid/.

[26] G. d'Atri and C. Puech, *Probabilistic analysis of the subset sum problem*, Discrete Appl. Math. 4 (1982), pp. 329–334.

[27] G. Debreu and I. Herstein, *Nonnegative square matrices*, Econometrica 21 (1953), pp. 597–607.

[28] P. Dehornoy, *Braid groups and left distributive operations*, Trans. Amer. Math. Soc. 345 (1994), pp. 115–151.

[29] _____, *A fast method for comparing braids*, Adv. Math. 125 (1997), pp. 200–235.

[30] _____, *Braids and Self Distributivity*. Birkhäuser, 2000.

[31] _____, *Braid-based cryptography*. Group theory, statistics, and cryptography, Contemporary Mathematics 360, pp. 5–33. American Mathematical Society, 2004.

[32] _____, *Using shifted conjugacy in braid-based cryptography*. Algebraic Methods in Cryptography, Contemporary Mathematics 418, pp. 65–74. American Mathematical Society, 2006.

[33] P. Dehornoy, I. Dynnikov, D. Rolfsen, and B. Wiest, *Why are braids orderable?*. Societe Mathematique De France, 2002.

[34] I. Dynnikov and B. Wiest, *On the Complexity of Braids*, preprint. Available at http://hal.archives-ouvertes.fr/hal-00001267/en/.

[35] T. ElGamal, *A Public-Key Cryptosystem and a Signature Scheme Based on Discrete Logarithms*, IEEE Trans. Inform. Theory IT-31 (1985), pp. 469–473.

[36] D. B. A. Epstein, J. W. Cannon, D. F. Holt, S. V. F. Levy, M. S. Paterson, and W. P. Thurston, *Word processing in groups*. Jones and Bartlett Publishers, 1992.

[37] U. Feige, A. Fiat, and A. Shamir, *Zero knowledge proofs of identity*, STOC '87: Proceedings of the nineteenth annual ACM conference on Theory of computing (1987), pp. 210–217.

[38] A. Fel'shtyn and E. Troitsky, *Twisted conjugacy separable groups*, preprint. Available at http://arxiv.org/abs/math/0606764.

[39] R. Fenn, M. T. Greene, D. Rolfsen, C. Rourke, and B. Wiest, *Ordering the braid groups*, Pacific J. Math. 191 (1999), pp. 49–74.

[40] N. Franco and J. González-Meneses, *Conjugacy problem for braid groups and Garside groups*, J. Algebra 266 (2003), pp. 112–132.

[41] D. Garber, *Braid group cryptography*, preprint. Available at http://arxiv.org/abs/0711.3941.

[42] D. Garber, S. Kaplan, M. Teicher, B. Tsaban, and U. Vishne, *Probabilistic solutions of equations in the braid group*, Adv. Appl. Math. 35 (2005), pp. 323–334.

[43] _____, *Length-based conjugacy search in the braid group.* Algebraic Methods in Cryptography, Contemporary Mathematics 418, pp. 75–88. American Mathematical Society, 2006.

[44] M. Garey and J. Johnson, *Computers and Intractability: A Guide to the Theory of NP-Completeness.* W. H. Freeman, 1979.

[45] P. Garrett, *Making, Breaking Codes: Introduction to Cryptology.* Prentice Hall, 2001.

[46] V. Gebhardt, *A New Approach to the Conjugacy Problem in Garside Groups,* J. Algebra 292 (2005), pp. 282–302.

[47] _____, *Conjugacy Search in Braid Groups From a Braid-based Cryptography Point of View,* Appl. Algebra Eng. Comm. 17 (2006), pp. 219–238.

[48] R. Gilman, A. G. Myasnikov, A. D. Miasnikov, and A. Ushakov, *Generic complexity of algorithmic problems,* Vestnik OMGU Special Issue (2007), pp. 103–110.

[49] O. Goldreich, *Foundations of cryptography.* Cambridge University Press, 2001.

[50] S. Goldwasser and S. Micali, *Probabilistic encryption,* J. Comput. Syst. Sci. 28 (1984), pp. 270–299.

[51] M. I. Gonzalez-Vasco, S. Magliveras, and R. Steinwandt, *Group Theoretic Cryptography.* Chapman & Hall/CRC, to appear in 2008.

[52] R. I. Grigorchuk, *Burnside's problem on periodic groups,* Funct. Anal. Appl. 14 (1980), pp. 41–43.

[53] _____, *On growth in group theory.* Proceedings of the International Congress of Mathematicians, pp. 325–338, Kyoto, 1990.

[54] _____, *On the problem of M. Day about nonelementary amenable groups in the class of finitely presented groups,* Math. Notes 60 (1996), pp. 774–775.

[55] _____, *On the system of defining relations and the Schur multiplier of periodic groups generated by finite automata.* GROUPS St Andrews 1997, pp. 290–317. Cambridge Univ. Press, 1999.

[56] D. Grigoriev and V. Shpilrain, *Zero-knowledge authentication schemes from actions on graphs, groups, or rings,* preprint. Available at http://arxiv.org/abs/0802.1661.

[57] N. Gupta, *Free Group Rings,* Contemporary Mathematics 66. American Mathematical Society, 1987.

[58] Y. Gurevich, *Average case completeness,* J. Comput. Syst. Sci. 42 (1991), pp. 346–398.

[59] _____, *From Invariants to Canonization.* Bulletin of the European Association for Theoretical Computer Science, pp. 327–331. World Scientific, 2001.

[60] _____, *The Challenger-Solver game: Variations on the theme of P =?NP.* Logic in Computer Science Column, The Bulletin of EATCS, pp. 112–121, October, 1989.

[61] Y. Gurevich and S. Shelah, *Expected computation time for Hamiltonian Path Problem*, SIAM J. Comput. 16 (1987), pp. 486–502.

[62] J. D. Hamkins and A. G. Myasnikov, *The halting problem is decidable on a set of asymptotic probability one*, Notre Dame Journal of Formal Logic 47 (2006), pp. 515–524.

[63] Pierre de la Harpe, *Topics in geometric group theory.* The University of Chicago Press, 2000.

[64] M. E. Hellman, *An Overview of Public Key Cryptography*, IEEE Communications Magazine (May 2002), pp. 42–49.

[65] D. Hofheinz and R. Steinwandt, *A practical attack on some braid group based cryptographic primitives.* Advances in Cryptology – PKC 2003, Lecture Notes in Computer Science 2567, pp. 187–198. Springer, Berlin, 2003.

[66] J. Hughes, *A linear algebraic attack on the AAFG1 braid group cryptosystem.* The 7th Australasian Conference on Information Security and Privacy ACISP 2002, Lecture Notes in Computer Science 2384, pp. 176–189. Springer, Berlin, 2002.

[67] J. Hughes and A. Tannenbaum, *Length-based attacks for certain group based encryption rewriting systems*, preprint. Available at http://front.math.ucdavis.edu/0306.6032.

[68] R. Impagliazzo, *A personal view of average-case complexity.* Proceedings of the 10th Annual Structure in Complexity Theory Conference (SCT'95), pp. 134–147, 1995.

[69] _____, *Computational Complexity Since 1980.* FSTTCS 2005: Foundations of Software Technology and Theoretical Computer Science, Lecture Notes in Computer Science 3821, pp. 19–47. Springer, Berlin, 2005.

[70] Clay Mathematical Institute, http://www.claymath.org/prizeproblems/pvsnp.htm.

[71] T. Jitsukawa, *Malnormal subgroups of free groups.* Computational and Statistical Group Theory, Contemporary Mathematics 298, pp. 83–96. American Mathematical Society, 2002.

[72] A. G. Kalka, *Representation Attacks on the Braid Diffie-Hellman Public Key Encryption*, Appl. Algebra Eng. Comm. 17 (2006), pp. 257–266.

[73] A. G. Kalka, M. Teicher, and B. Tsaban, *Cryptanalysis of the algebraic eraser*, in preparation.

[74] I. Kapovich and A. G. Miasnikov, *Stallings foldings and subgroups of free groups*, J. Algebra 248 (2002), pp. 608–668.

[75] I. Kapovich, A. G. Miasnikov, P. Schupp, and V. Shpilrain, *Generic-case complexity, decision problems in group theory and random walks*, J. Algebra 264 (2003), pp. 665–694.

[76] _____, *Average-case complexity and decision problems in group theory*, Adv. Math. 190 (2005), pp. 343–359.

[77] I. Kapovich and P. Schupp, *Genericity, the Arzhantseva-Ol'shanskii method and the isomorphism problem for one-relator groups*, Math. Ann. 331 (2005), pp. 1–19.

[78] I. Kapovich, P. Schupp, and V. Shpilrain, *Generic properties of Whitehead's algorithm and isomorphism rigidity of random one-relator groups*, Pacific J. Math. 223 (2006), pp. 113–140.

[79] H. Kellerer, U. Pferschy, and D. Pisinger, *Knapsack Problems*. Springer, 2004.

[80] L. Khachiyan, *A polynomial algorithm in linear programming*, Dokl. Akad. Nauk SSSR 244 (1979), pp. 1093–1096.

[81] O. Kharlampovich and M. Sapir, *Algorithmic problems in varieties*, Internat. J. Algebra and Comput. 5 (1995), pp. 379–602.

[82] V. Klee and G. Minty, *How good is the simplex algorithm?*. Inequalities, III (Proc. Third Sympos., Univ. California, Los Angeles, Calif., 1969), pp. 159–175. Academic Press, 1972.

[83] D. Knuth, J. H. Morris, and V. Pratt, *Fast pattern matching in strings*, SIAM J. Comput. 6 (1977), pp. 323–350.

[84] K. H. Ko, S. J. Lee, J. H. Cheon, J. W. Han, J. Kang, and C. Park, *New public-key cryptosystem using braid groups*. Advances in Cryptology – CRYPTO 2000, Lecture Notes in Computer Science 1880, pp. 166–183. Springer, Berlin, 2000.

[85] D. Krammer, *Braid groups are linear*, Ann. Math. 155 (2002), pp. 131–156.

[86] Y. Kurt, *A New Key Exchange Primitive Based on the Triple Decomposition Problem*, preprint. Available at `http://eprint.iacr.org/2006/378`.

[87] J. C. Lagarias, *The $3x + 1$ problem and its generalizations*, Amer. Math. Monthly 92 (1985), pp. 3–23.

[88] S. Lal and A. Chaturvedi, *Authentication Schemes Using Braid Groups*, preprinet. Available at `http://arxiv.org/abs/cs/0507066`, 2005.

[89] E. Lee, *Right-Invariance: A Property for Probabilistic Analysis of Cryptography based on Infinite Groups.* Advances in Cryptology – Asiacrypt 2004, Lecture Notes in Computer Science 3329, pp. 103–118. Springer, Berlin, 2004.

[90] S. J. Lee, *Algorithmic Solutions to Decision Problems in the Braid Group,* Ph.D. thesis, KAIST, 2000.

[91] S. J. Lee and E. Lee, *Conjugacy classes of periodic braids,* preprint. Available at http://front.math.ucdavis.edu/0702.5349.

[92] _____ , *Potential Weaknesses of the Commutator Key Agreement Protocol Based on Braid Groups.* Advances in Cryptology – EUROCRYPT 2002, Lecture Notes in Computer Science 2332, pp. 14–28. Springer, Berlin, 2002.

[93] L. Levin, *Average case complete problems,* SIAM J. Comput. 15 (1986), pp. 285–286.

[94] M. Li and P. Vitanyi, *An Introduction to Kolmogorov Complexity and Its Applications,* Graduate texts in Computer Science. Springer, 1997.

[95] R. Lyndon and P. Schupp, *Combinatorial Group Theory,* Classics in Mathematics. Springer-Verlag, 2001.

[96] W. Magnus, A. Karrass, and D. Solitar, *Combinatorial Group Theory.* Springer-Verlag, 1977.

[97] M. R. Magyarik and N. R. Wagner, *A Public Key Cryptosystem Based on the Word Problem.* Advances in Cryptology – CRYPTO 1984, Lecture Notes in Computer Science 196, pp. 19–36. Springer, Berlin, 1985.

[98] A. Martino, E. Turner, and E. Ventura, *Counting monomorphisms of a free group,* in preparation.

[99] F. Matucci, *Cryptanalysis of the Shpilrain-Ushakov Protocol for Thompson's Group,* To appear in Journal of Cryptology.

[100] A. J. Menezes, *Handbook of Applied Cryptography.* CRC Press, 1996.

[101] A. G. Miasnikov and D. Osin, *Random subgroups of finitely generated groups,* in preparation.

[102] A. G. Miasnikov, V. Shpilrain, and A. Ushakov, *A practical attack on some braid group based cryptographic protocols.* Advances in Cryptology – CRYPTO 2005, Lecture Notes in Computer Science 3621, pp. 86–96. Springer, Berlin, 2005.

[103] _____ , *Random Subgroups of Braid Groups: An Approach to Cryptanalysis of a Braid Group Based Cryptographic Protocol.* Advances in Cryptology – PKC 2006, Lecture Notes in Computer Science 3958, pp. 302–314. Springer, Berlin, 2006.

[104] A. G. Miasnikov and A. Ushakov, *Generic complexity of the conjugacy search problem in groups*, in preparation.

[105] _____, *Random van Kampen diagrams and the word problem in groups*, in preparation.

[106] _____, *Random subgroups and analysis of the length-based and quotient attacks*, J. Math. Cryptology 2 (2008), pp. 29–61.

[107] K. A. Mihailova, *The occurrence problem for direct products of groups*, Dokl. Akad. Nauk SSSR 119 (1958), pp. 1103–1105.

[108] C. F. Miller III, *Decision problems for groups – survey and reflections.* Algorithms and Classification in Combinatorial Group Theory, pp. 1–60. Springer, 1992.

[109] A. D. Myasnikov and A. G. Miasnikov, *Whitehead method and Genetic Algorithms.* Computatyional and experimental group theory, Contemporary Mathematics 349, pp. 89–114. American Mathematical Society, 2004.

[110] A. D. Myasnikov, A. G. Miasnikov, and V. Shpilrain, *On the Andrews-Curtis equivalence.* Combinatorial and geometric group theory, Contemporary Mathematics 296, pp. 183–198. American Mathematical Society, 2002.

[111] A. D. Myasnikov and A. Ushakov, *Cryptanalysis of the Anshel-Anshel-Goldfeld-Lemieux key agreement protocol*, to appear in Groups, Complexity, and Cryptology.

[112] A. D. Myasnikov and A. Ushakov, *Length Based Attack and Braid Groups: Cryptanalysis of Anshel-Anshel-Goldfeld Key Exchange Protocol.* Advances in Cryptology – PKC 2007, Lecture Notes in Computer Science 4450, pp. 76–88. Springer, Berlin, 2007.

[113] D. Osin and V. Shpilrain, *Public key encryption and encryption emulation attacks.* Computer Science in Russia 2008, Lecture Notes in Computer Science 5010, pp. 252–260. Springer, 2008.

[114] C. Papadimitriou, *Computational Complexity.* Addison-Wesley, 1994.

[115] M. Paterson and A. Razborov, *The set of minimal braids is co-NP-complete*, J. Algorithms 12 (1991), pp. 393–408.

[116] D. Peifer, *Artin groups of extra-large type are automatic*, J. Pure Appl. Algebra 110 (1996), pp. 15–56.

[117] G. Petrides, *Cryptanalysis of the Public Key Cryptosystem Based on the Word Problem on the Grigorchuk Groups.* 9th IMA International Conference on Cryptography and Coding, Lecture Notes in Computer Science 2898, pp. 234–244. Springer, Berlin, 2003.

[118] D. Ruinsky, A. Shamir, and B. Tsaban, *Cryptanalysis of group-based key agreement protocols using subgroup distance functions.* Advances in Cryptology – PKC 2007, Lecture Notes in Computer Science 4450, pp. 61–75. Springer, Berlin, 2007.

[119] V. Shpilrain, *Assessing security of some group based cryptosystems.* Group theory, statistics, and cryptography, Contemporary Mathematics 360, pp. 167–177. American Mathematical Society, 2004.

[120] _____, *Hashing with polynomials.* Information Security and Cryptology – ICISC 2006, Lecture Notes in Computer Science 4296, pp. 22–28. Springer, 2006.

[121] _____, *Cryptanalysis of Stickel's key exchange scheme.* Computer Science in Russia 2008, Lecture Notes in Computer Science 5010, pp. 283–288. Springer, 2008.

[122] V. Shpilrain and A. Ushakov, *Thompson's group and public key cryptography.* Applied Cryptography and Network Security – ACNS 2005, Lecture Notes in Computer Science 3531, pp. 151–164. Springer, 2005.

[123] _____, *The conjugacy search problem in public key cryptography: unnecessary and insufficient,* Appl. Algebra Engrg. Comm. Comput. 17 (2006), pp. 285–289.

[124] _____, *A new key exchange protocol based on the decomposition problem.* Algebraic Methods in Cryptography, Contemporary Mathematics 418, pp. 161–167. American Mathematical Society, 2006.

[125] V. Shpilrain and G. Zapata, *Using decision problems in public key cryptography,* preprint. Available at
http://www.sci.ccny.cuny.edu/ shpil/res.html.

[126] _____, *Combinatorial group theory and public key cryptography,* Appl. Algebra Engrg. Comm. Comput. 17 (2006), pp. 291–302.

[127] _____, *Using the subgroup membership search problem in public key cryptography.* Algebraic Methods in Cryptography, Contemporary Mathematics 418, pp. 169–179. American Mathematical Society, 2006.

[128] H. Sibert, P. Dehornoy, and M. Girault, *Entity authentication schemes using braid word reduction,* Discrete Appl. Math. 154 (2006), pp. 420–436.

[129] V. M. Sidelnikov, M. A. Cherepnev, and V. Y. Yashcenko, *Systems of open distribution of keys on the basis of noncommutative semigroups,* Russian Acad. Sci. Dokl. Math. 48 (1994), pp. 384–386.

[130] S. Smale, *On the average number of steps of the simplex method of linear programming,* Math. Program. 27 (1983), pp. 241–262.

[131] M. Sramka, *On the Security of Stickel's Key Exchange Scheme,* preprint.

[132] J. Stallings, *Topology of finite graphs*, Invent. Math. 71 (1983), pp. 551–565.

[133] E. Stickel, *A New Method for Exchanging Secret Keys*. Proceedings of the Third International Conference on Information Technology and Applications (ICITA05), Contemporary Mathematics 2, pp. 426–430. IEEE Computer Society, 2005.

[134] D. R. Stinson, *Cryptography: Theory and Practice*, Discrete Mathematics and Its Applications. Chapman & Hall/CRC, 2005.

[135] J. Talbot and D. Welsh, *Complexity and Cryptography: An Introduction*. Cambridge University Press, 2006.

[136] J.-P. Tillich and G. Zémor, *Hashing with $SL_2$*. Advances in Cryptology – CRYPTO 1994, Lecture Notes in Computer Science 839, pp. 40–49. Springer, 1994.

[137] N. Touikan, *A Fast Algorithm for Stallings' Folding Process*, Internat. J. Algebra and Comput. 16 (2006), pp. 1031–1046.

[138] A. Ushakov, *Fundamental Search Problems in Groups*, Ph.D. thesis, CUNY/Graduate Center, 2005.

[139] A. M. Vershik, S. Nechaev, and R. Bikbov, *Statistical properties of braid groups with application to braid groups and growth of heaps*, Commun. Math. Phys. 212 (2000), pp. 469–501.

[140] A. M. Vershik and P. V. Sporyshev, *An estimate of the average number of steps in the simplex method, and problems in asymptotic integral geometry*, Dokl. Akad. Nauk SSSR 271 (1983), pp. 1044–1048.

[141] W. Woess, *Cogrowth of groups and simple random walks*, Arch. Math. 41 (1983), pp. 363–370.

# Abbreviations and Notation

# Index

# Advanced Courses in Mathematics CRM Barcelona

Edited by
**Manuel Castellet**

Since 1995 the Centre de Recerca Matemàtica (CRM) in Barcelona has conducted a number of annual Summer Schools at the post-doctoral or advanced graduate level. Sponsored mainly by the European Community, these Advanced Courses have usually been held at the CRM in Bellaterra.
The books in this series consist essentially of the expanded and embellished material presented by the authors in their lectures.

**Argyros, S. / Todorcevic, S.**
Ramsey Methods in Analysis (2005)
ISBN 978-3-7643-7264-4

This book introduces graduate students and researhers to the study of the geometry of Banach spaces using combinatorial methods. We show, for example, how to introduce a conditional structure to a given Banach space under construction that allows us to essentially prescribe the corresponding space of non-strictly singular operators. We also apply the Nash-Williams theory of fronts and barriers in the study of Cezaro summability and unconditionality present in basic sequences inside a given Banach space. We further provide a detailed exposition of the block-Ramsey theory and its recent deep adjustments relevant to the Banach space theory due to Gowers.

**Audin, M. / Cannas da Silva, A. /
Lerman, E.**
Symplectic Geometry of Integrable Hamiltonian Systems (2003)
ISBN 978-3-7643-2167-3

**Brady, N. / Riley, T. / Short, H.**
The Geometry of the Word Problem for Finitely Generated Groups (2006)
ISBN 978-3-7643-7949-0

The origins of the word problem are in group theory, decidability and complexity, but, through the vision of Gromov and the language of filling functions, the topic now impacts the world of large-scale geometry, including topics such as soap films, isoperimetry, coarse invariants and curvature.
The first part introduces van Kampen diagrams in Cayley graphs of finitely generated, infinite groups; it discusses the van Kampen lemma, the isoperimetric functions or Dehn functions, the theory of small cancellation groups and an introduction to hyperbolic groups. The second part is dedicated to Dehn functions, negatively curved groups, in particular, CAT(0) groups, cubings and cubical complexes. In the last part, filling functions are presented from geometric, algebraic and algorithmic points of view. Many examples and open problems are included.

**Brown, K.A. / Goodearl, K.R.**
Lectures on Algebraic Quantum Groups (2002)
ISBN 978-3-7643-6714-5

**Catalano, D. / Cramer, R. / Damgård, I. /
Di Creszenso, G. / Pointcheval, D. / Takagi, T.**
Contemporary Cryptology (2005)
ISBN 978-3-7643-7294-1

**Christopher, C. / Li, C.**
Limit Cycles of Differential Equations (2007)
ISBN 978-3-7643-8409-8

**Cohen, R.L. / Hess, K. / Voronov, A.A.**
String Topology and Cyclic Homology (2006)
ISBN 978-3-7643-2182-6

The subject of this book is string topology, Hochschild and cyclic homology. The first part consists of an excellent exposition of various approaches to string topology and the Chas-Sullivan loop product. The second gives a complete and clear construction of an algebraic model for computing topological cyclic homology. The book provides many references for the reader wishing to learn more about the subject, to which it gives a perfect introduction. It is therefore suitable for both graduate students and established researchers. It is certainly the best source of much information that was until now available only to specialists and covers material from the elementary bases to the most recent developments.

**Da Prato, G.**
Kolmogorov Equations for Stochastic PDEs (2004)
ISBN 978-3-7643-7216-3

**Drensky, V. / Formanek, E.**
Polynomial Identity Rings (2004)
ISBN 978-3-7643-7126-5

**Dwyer, W.G. / Henn, H.-W.**
Homotopy Theoretic Methods in Group Cohomology (2001)
ISBN 978-3-7643-6605-6

**Markvorsen, S. / Min-Oo, M.**
Global Riemannian Geometry: Curvature and Topology (2003)
ISBN 978-3-7643-2170-3

**Mislin, G. / Valette, A.**
Proper Group Actions and the Baum-Connes Conjecture (2003)
ISBN 978-3-7643-0408-9

BIRKHÄUSER